Chemical
Pattern Recognition

CHEMOMETRICS SERIES

Series Editor: **Dr. D. Bawden**
Pfizer Central Research, Sandwich, Kent, England

Chemical
Pattern Recognition

O. Štrouf

Czechoslovak Academy of Sciences, Czechoslovakia

RESEARCH STUDIES PRESS LTD.
Letchworth, Hertfordshire, England

JOHN WILEY & SONS INC.
New York · Chichester · Toronto · Brisbane · Singapore

RESEARCH STUDIES PRESS LTD.
58B Station Road, Letchworth, Herts. SG6 3BE, England

Marketing and Distribution:

Australia, New Zealand, South-east Asia:
Jacaranda-Wiley Ltd., Jacaranda Press
JOHN WILEY & SONS INC.
GPO Box 859, Brisbane, Queensland 4001, Australia

Canada:
JOHN WILEY & SONS CANADA LIMITED
22 Worcester Road, Rexdale, Ontario, Canada

Europe, Africa:
JOHN WILEY & SONS LIMITED
Baffins Lane, Chichester, West Sussex, England

North and South America and the rest of the world:
JOHN WILEY & SONS INC.
605 Third Avenue, New York, NY 10158, USA

Library of Congress Cataloging in Publication Data

Štrouf, Oldřich.
  Chemical pattern recognition.

  (Chemometrics series; 11)
  Bibliography: p.
  Includes index.
    1. Chemistry, Analytic——Data processing.   2. Optical
pattern recognition.   I. Title.   II. Series.
QD75.4.E4S77   1986   543   86-15956
ISBN 0 86380 044 0
ISBN 0 471 91252 2 (Wiley)

British Library Cataloguing in Publication Data

Štrouf, O.
  Chemical pattern recognition.——
  (Chemometrics series; 11)
  1. Chemistry, Analytic——Statistical
  methods   2. Multivariate analysis
  3. Pattern perception
  I. Title   II. Series
  543'.001'519535   QD75.4.S8

  ISBN 0 86380 044 0
  ISBN 0 471 91252 2 Wiley

  ISBN 0 86380 044 0 (Research Studies Press Ltd.)
  ISBN 0 471 91252 2 (John Wiley & Sons Inc.)

Printed in Great Britain by Short Run Press Ltd., Exeter

To my wife Alexandra

Editorial Preface

One of the problems of newly emergent scientific disciplines is that relevant information may be widely scattered across a diverse literature. The writing of reviews, and the compilation of bibliographies, greatly assists the progress of rapidly developing fields of study. Conversely, the appearance of such literature surveys may be viewed as a sign that a coherent discipline, or sub-discipline, is forming.

At the beginning of this book, Dr. Štrouf notes that the application of pattern recognition in chemistry is a distinct subject area now showing signs of reaching maturity. The appearance of this book is a clear token that this is indeed the case. Dr. Štrouf's compilation of over 300 references, all to relatively recent work, attests the widespread research interest in the subject. This clear and thoughtful survey of the significance of, and inter-relationships between, the various topics included will be of great value to anyone interested in assessing the progress of this young discipline.

David Bawden
Sandwich
January 1986

Preface

Recognition of patterns is a natural process of human perception of the objective world. A life-long training results in the extraordinary ability of human beings to classify patterns, and they are the best pattern recognizer known today.

The human recognition ability is, nevertheless, restricted to three-dimensional space only. But many practical as well as scientific problems are inherently multidimensional. Therefore, in the analysis of such problems mathematical tools should be used, because mathematics does not suffer from the human low-dimension limitation.

For mathematical pattern recognition, the j-th phenomenon (object) of the studied system is represented by a data vector, pattern, $\underline{x} = (x_1, x_2, \ldots, x_i, \ldots, x_N)$ where x_i is the value for the variable i and N is the number of the variables (in multidimensional cases $N > 3$). The variables form an N-dimensional measurement space in which pattern $\underline{x}$ can be geometrically represented as a point with coordinates $x_1, x_2, \ldots, x_i, \ldots, x_N$ in this space. The low-dimensional representation is illustrative and will be used for graphical demonstration, $1 \leqq N \leqq 3$ usually 2, in this book; principles valid for these graphs remain valid for multidimensional problems.

A multidimensional pattern characterized by such a data string can be used to define some required property of the objects. This requires much computation, which is a reason why the mathematical pattern recognition has been developing in connection with the construction of highly efficient digital computers with large memory. The use of computers in analyses of systems with multidimensional patterns is expressed by the often used terms "computerized pattern recognition" or "computer-aided (assisted) pattern recognition".

The aim of many mathematical computerized pattern recognition methods (termed throughout this book simply "pattern recognition") is to reduce the dimensionality of the system to the low-dimensional representation suitable for human recognition.

The basic purpose of pattern recognition is to classify the objects of the system into subsets of objects with the same type of property, here called "classification property"; the objects with similar classification property form classes (categories). There is an enormous variety of possible approaches to the solution of this problem. The tools are borrowed from many other disciplines such as informatics, probability theory, statistical decision theory, threshold logic, dynamic programming and many others. This variety results only in rather fuzzy definitions of this methodology (Verhagen, 1975). The extremely concise definition has been recently formulated by Verhagen et al. (1980), in which pattern recognition is considered as a many-to-one mapping.

In spite of the lack of a rigid definition, the pattern recognition methodology has been developed and used in many scientific and especially practical activities. The main reason is evidently the fact that mankind is daily compelled to solve many dramatically important

problems, for which an exact theoretical solutions is not
available. As examples, the use of pattern recognition
in environmental problems, in material prospecting
problems, in medicine etc. can be noted.

Moreover, there exist many cases in which a pattern
recognition method is used in the automation of a certain
process performed up to now by humans. Some features of
the automation process using pattern recognition simulate
to some extent human intelligence and, therefore, it
often overlaps with artificial intelligence.

Along with this "practical" use of pattern recognition,
the possibility to analyse the behaviour of a system is
also useful.

Pattern recognition was introduced in chemistry in the
late sixties. During the past years it has been demonstrated that chemical pattern recognition can be valuable
in solving many and various problems, if carefully used.
It should be emphasized here that the chemical pattern
recognition in this book is considered as pattern recognition analyses of chemical data irrespective of the
type of the classification property. If the classification property is a chemical property, a "pure" chemical
pattern recognition problem arises (e.g. the features x_i
are spectral data and the classification property is some
structural information). On the other hand, a set of
chemical features measured on the objects (phenomena) of
real world can be used to classify these objects
(phenomena) according to some   non-chemical property. Such a non-chemical classification property can
be a physical property (e.g. superconductivity) or a biological property (e.g. carcinogenicity) or other types
of properties. In this author's opinion, these chemical
pattern recognition applications in non-chemical problems
can be highly valuable.

Particularly, attention has been recently focused on automation of processing large sets of chemical data as produced at an increasing rate by efficient (and expensive) analytical methods (e.g. gas chromatographic techniques combined with mass spectrometric or vapour-phase infrared spectral methods). It is also commonly known that routine analyses, for example in quality control and biochemical testing, also yield large collections of data. The chemical information processing of such data sets by pattern recognition is discussed in Section 1.1.

Moreover, pattern recognition may be suitable as one of possible approaches to modelling some types of chemical systems (Section 1.2).

If the variables selected for the characterization of the objects possess a fundamental physical meaning, the modelling of chemical systems could open discussion about the possible use of the results for the critical assessment and innovation of some theoretical aspects.

The high applicability of chemical pattern recognition is one of the reasons for the very broad interest in this approach. The progress up to 1975 has been summarized in the first monograph on chemical pattern recognition written by Jurs and Isenhour (1975). Another monograph of Varmuza (1980) describes the progress up to 1980. Here, an attempt is made to summarize and, to some extent, discuss recent results achieved from 1979 up to 1985. Recent publications are discussed in some detail, while only selected works published before 1979 are involved because of the availability of comprehensive references in the above mentioned excellent books.

Only a few basic methodological approaches (Chapter 2) to chemical patterns classification (Sections 2.1 and 2.2), feature selection (Section 2.3) and visualization (Section 2.4) are outlined here without detailed

mathematical description. Recent applications of pattern
recognition are summarized in Chapter 3 of the monograph
for different types of objects (Section 3.1) and various
classification properties (Section 3.2).

The results treated in this book could serve, in our
opinion, for an interested reader as informative material
for a critical assessment of recent trends in chemical
pattern recognition. In this connection, the other related books and reviews (see Chapter 1) have to be con-
sulted. Here, the fourth Chapter of the monograph can only
make an attempt to pick up some more significant trends
as they may be extracted from the recent published
information (see References).

Nowadays,   chemical pattern recognition represents
one of the fundamental parts of chemometrics (Kowalski,
1980; Frank and Kowalski, 1982; Delaney, 1984), a sub-
discipline of chemistry founded in 1974 by Kowalski and
Wold (Kowalski, 1975, 1977,p.VII). Therefore, it should
be recommended to consider also the publications dealing
with chemometrics or its parts. In this connection it is
worth mentioning that pattern recognition uses and
sophistically combines other chemometrical disciplines
of a "lower level" for classification purposes. Thus,
very valuable information may be found in these individual disciplines.

The author would like to acknowledge Professor Danzer
(Jena University), Professor Isenhour (Utah State University), Professor Jurs (Pennsylvania State University),
Professor Massart (Vrije University Brussel), Professor
Varmuza (Technical University Vienna) and Professor Wold
(Umeå University) for their kind offering of recent
informative materials about fundamental developments
achieved by them in chemical pattern recognition
including those prior to publication. The author would like
to thank Dr. Fusek for discussion about the SPHERE

algorithm and assistance in conducting the illustrative
material and Dr. Kuchynka for discussions about some
philosophical aspects of chemical pattern recognition.
The author is particularly indebted to Dr. Eckschlager
for encouraging   comments on the manuscript and to
Dr. Štěpánek for comments on the manuscript and for
language revision.

Last but not least, the author would like to thank
Dr. Bawden, the Editor of the Chemometrics Research
Studies Series, for his stimulation to write this mono-
graph and his assistance in its preparation.

Prague, March 1985

O. Štrouf

Contents

CHAPTER 1

Introduction to Chemical Pattern Recognition

The history of each scientific discipline can be used, in a critical way, for indication of trends in the near future. Furthermore, one has to bear in mind that no discipline develops independently and that each is strongly influenced by the development of other disciplines. As mentioned in the Preface, the progress of pattern recognition is closely connected with the construction of more and more effective and available computers.

The computerized pattern recognition was used primarily for automation of information processing in the early fifties (see e.g. Tou and Gonzales, 1974, Russian translation, p.8). In the late fifties similar features of pattern recognition and human perception were studied under the name of perceptrons (Rosenblatt, 1962,1964).

An explosion of pattern recognition applications in many various branches could be observed in 1960-1980 and in chemistry in the 1969-1980 period; in the latter case more than 360 publications on pattern recognition in chemistry have been published (Varmuza, 1980,pp.190-200). An over-optimism was natural for this stage, because insufficiently critical expectations are characteristic for each discipline in its infancy before the determination of its scope and limitation.

Therefore, a stage of maturing of pattern recognition can (and should) be expected in the eighties. Actually, a more critical approach to chemical pattern recognition can be recognized in some of recent sources of information which are available in monographs and in chapters of books (Bailey, 1979; Eckschlager and Štěpánek, 1979; Hodes, 1979; Jurs et al., 1979a, 1983a; Kirschner and Kowalski, 1979; Mellon, 1979; Petit et al., 1979; Stuper et al., 1979; Golender and Rozenblit, 1980; Lewi, 1980; Malinowski and Howery, 1980; Sjöström and Wold, 1980; Varmuza, 1980, 1984a; Kateman and Pijpers, 1981; Justice and Isenhour,Eds., 1982, 1982a; Kowalski and Wold, 1982; Zupan, 1982a; Eckschlager et al., 1983; Heller and Potenzone,Eds., 1983; Massart and Kaufman, 1983a; Jurs, 1984; Pungor et al.,Eds., 1984).

Moreover, chemical pattern recognition or some of its special parts are summarized in a series of reviews and surveys (Coomans et al., 1979; Massart and Michotte, 1979; McLafferty and Venkataragharan, 1979; Varmuza, 1979, 1980a, 1982; Kowalski, 1980, 1981; Kryger, 1980, 1981; Veress, 1982; Jellum, 1981a; Martinsen, 1981; Miyashita et al., 1981c, 1982; Whalen-Pedersen and Jurs, 1981; Bisani et al., 1982, 1982a; Derde and Massart, 1982; Frank and Kowalski, 1982; Massart, 1982; Massart et al., 1982a; Ioffe et al., 1983; Ordukhanyan et al., 1983; Voorhees et al., 1983; Delaney, 1984; Rusling, (1984).

Chemical pattern recognition is involved also in symposium proceedings (Ioffe et al., 1980; Fusek and Štrouf, 1980; Štrouf, 1980; Bertsch et al., 1981; Massart, 1981a; Kuchynka et al., 1982; Arunachalam et al., 1983; Dunn and Wold, 1983; Meglen and Erickson, 1983; Wold et al., 1983a, 1983b, 1984, 1984a; Varmuza, 1983, 1984b; Varmuza and Lohninger, 1984; Isaszegi-Vass et al., 1984; Veress and Pungor, 1984), in reports (Varmuza, 1981; Dunn et al., 1982; Meuzelaar, 1982; Perone et al., 1982; Wold et al., 1982) and in dissertations (Schachterle, 1980; Byers, 1981; Tsao, 1982b; Donald, 1984).

A pattern recognition system can be generally considered as a mathematical construction of classification using more or less heuristic connections (see Chapter 2). In practice, pattern recognition starts with the reception of signals from real world by means of a sensor followed by the processing of these signals to an acceptable form for computers, and it ends with the representation of classification results in a form compatible for humans, preferably by visualization. The whole system is multistep in its nature and can be of relatively high complexity with interrelations between the individual steps (for a simplified scheme see Fig.1).

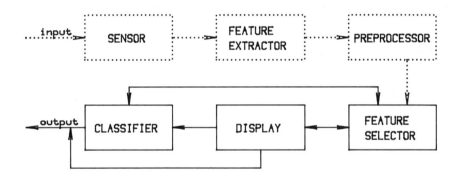

Fig.1. Pattern recognition system.
Fully drawn parts correspond to the pattern-recognition parts discussed in this monograph; searching for the optimum combination of these parts can be considered as an "art of pattern recognition".

In this monograph, only the principal parts of the pattern recognition system are discussed, namely classification (Sections 2.1 and 2.2), feature selection (Section 2.3) and visualization (Section 2.4). The other parts are procedures used commonly in other multivariate data analyses and they are described elsewhere in detail. Hencefore, they are only mentioned in Section 3.1 in connection with preprocessing of chemical data.

As a start here, extracted and preprocessed features in a digital form, x_i ($i = 1,2,...,N$), are considered for the patterns $\underline{x}_j$ ($j = 1,2,...,M$, where M is the number of patterns). In other words, a table (matrix) of data stands at the beginning (Fig.2).

	x_1	x_2	$\cdot$	x_i	$\cdot$	x_N
$\underline{x}_1$	x_{11}	x_{21}	$\cdot$	x_{i1}	$\cdot$	x_{N1}
$\underline{x}_2$	x_{12}	x_{22}	$\cdot$	x_{i2}	$\cdot$	x_{N2}
$\cdot$	$\cdot$	$\cdot$	$\cdot$	$\cdot$	$\cdot$	$\cdot$
$\underline{x}_j$	x_{1j}	x_{2j}	$\cdot$	x_{ij}	$\cdot$	x_{Nj}
$\cdot$	$\cdot$	$\cdot$	$\cdot$	$\cdot$	$\cdot$	$\cdot$
$\underline{x}_M$	x_{1M}	x_{2M}	$\cdot$	x_{iM}	$\cdot$	x_{NM}

Fig.2. Table of data.
Columns represent the features x_i and rows the patterns $\underline{x}_j$; the elements x_{ij} are data.

The table of data can be, generally, viewed as a starting point in multivariate data analysis, pattern recognition being one of the most sophisticated of them. The aim is to convert a data table to a few informative pictures showing the relations between objects $\underline{x}_j$ (table rows) and variables x_i (table columns). Thus the important fundamental goals are searching for regularities in the tabulated data (see also Wold et al., 1984a) in respect to (a) feature selection according to their classification relevance and (b) classification of the objects using these features. The regularities found can be then used for two main objectives of the chemical pattern recognition:

(1) for the processing of chemical information contained in large collection of data (Section 1.1) and
(2) for the chemical modelling of the system under consideration (Section 1.2).

1.1 CHEMICAL INFORMATION PROCESSING BY PATTERN RECOGNITION

Pattern recognition represents a chemometric technique useful in obtaining information from large chemical data sets which cannot be extracted by human interpreter at all or, at least, in such an efficient and reliable way. Pattern recognition analysis can thus make the most of the large data sets generated in large-scale, and expensive, scientific programmes.

Pattern recognition principles have been applied in sophisticated automated structure elucidation (classification) systems. Computer-aided structure elucidation systems consist of three major processes: (1) partial-structure elucidation, (2) structure generation and (3) structure examination; and they must fulfil the following conditions: (a) reliability, (b) generality of application, (c) ease of modification and (d) portability (Abe et al., 1981).

Several hierarchical cluster analysis techniques for automatic classification of chemical structures give

different results as shown by Adamson and Bawden (1981).
Willett has studied methods of searching in chemical
structure files using intermolecular similarity coefficients (1982), different hierarchical clustering
algorithms (1982a) and some heuristics for nearest
neighbour searching (1983).

A series of publications is concerned with the information extractable from spectral data collections. The
data retrieval and artificial intelligence systems for
identifying polyatomic molecules from their molecular
spectra have been considered by Gribov (1980) as a
promising development for the future of molecular spectroscopy. Farkas et al. (1980, 1981) have described a
hierarchically organized computerized analytical system
ASSIGNER for determination of probability of the
presence of specific functional groups using on-line
measured spectral data and off-line collected parameters.
Pattern recognition techniques have been applied by
Frankel (1984) to the analysis of the library of Fourier
transformed infrared spectra. Recently, a new clustering
method suitable for large infrared spectral data sets has
been developed by Zupan (1982, 1982a) (see Sub-section
2.2.1).

The identification of mass spectra can be carried out
by retrieval systems such as, e.g., the PEAK system or
the Self-Training Interpretative and Retrieval System
(STIRS) (McLafferty and Venkataraghavan, 1979). Decision
-tree pattern recognition is recommended by Meisel et al.
(1979) for computer-aided identification of mass spectra
in large data bases.

Improvements of spectral library searches have been
systematically studied by Isenhour and his co-workers for
infrared spectra (Rasmussen and Isenhour, 1979b; Small
et al., 1979; Hangac et al., 1982; Owens and Isenhour,
1983) and for mass spectra (Rasmussen and Isenhour, 1979d;

Rasmussen et al., 1979, 1979a, 1979c; Lam R.B. et al., 1981; Williams et al., 1983). Kwiatkowski and Rieppe (1984) have shown that the success of spectral information selection from a spectral library by pattern recognition or retrieval methods depends on the quality of the training set and that it could be optimized by suitable transformation procedures and by using the means of information theory.

The processing of biochemical data is also of importance. The extraction of information from large sets of biochemical data (e.g. from clinical diagnosis, food analysis and pharmacology) by pattern recognition has been critically discussed by Derde and Massart (1982). Processing of metabolic profiles from biomedical gas chromatography—mass spectrometry data has been reviewed by Jellum (1981a). Hodes (1981, 1981a) has used a computer-aided statistical-heuristic method for processing a large set of antitumour prescreen data measured at the National Cancer Institute in the U.S.A. (more than 10 000 compounds per year).

Moreover, it should be emphasized that the large data sets do not serve only for pattern recognition extraction of information about the objects under consideration, but they can be employed, on the other hand, for objective assessment of the used analytical methods as described by Jansen et al. (1981, 1981a) for routine analytical methods in clinical laboratories.

Recently, the problem of the automated extraction of information from large data sets by pattern recognition methods has been critically discussed by Massart (1982).

1.2 MODELLING OF CHEMICAL SYSTEMS BY PATTERN RECOGNITION

Whereas in   pattern recognition data processing the
aim is to extract information from a given set of data
(collection, file), the goal of modelling by pattern recognition is to assess if artificially formed sets of
features are sufficiently representative for the classification purposes. In the latter case the patterns are
composed, on the basis of a priori knowledge, of features
hopefully relevant to the modelling of the studied system.
It should be stressed here, that under the term "models"
two different models are considered (in the text they are
not specified when the type is clear from the context):

(a)   chemical models of the objects as discussed in this
      Section and
(b)   mathematical models of the classes as treated in Sub
      -section 2.1.3.

In chemistry, analogously to other sciences, the
modelling of systems is used to search for regularities
in their behaviour with the aim of prediction. The level
of inherent complexity of chemical systems varies from
relatively simple systems to highly complex ones and,
therefore, general recipes for chemical modelling cannot
exist.

Attempts have been made to systemize chemical models
according to many different criteria. The fundamental
categorization of the models was described byaSwedish
chemometric school which differentiates the chemical
models by means of two very general criteria (Wold and
Sjöström, 1977):

(a)   the first criterion categorizes the models according
      to the range of their validity over the system;
      those generally valid are named global, the

remaining ones local,

(b)   the second criterion divides the models according
        to the character and accuracy of data characterizing
        the objects of the system;
        fundamental and accurate data enable the use of hard
        models, whereas derived and diffuse data require
        soft models.

The extreme cases in this categorization are thus represented by "hard-global" models valid mainly in physics
and physical chemistry for moderately complex systems
and, on the other hand, by "soft-local" models useful for
rather complicated chemical and especially biological
systems. For the latter ones the concept of similarity is
highly valuable (Wold and Sjöström, 1977).

Therefore, pattern recognition can be a suitable
methodological approach to "soft-local" modelling, because
it categorizes objects of a system according to their
similarity in respect of a classification property; the
similarity can be estimated by the analysis of the set of
multivariate data of the objects, patterns $\underline{x}$ (Štrouf,
1980).

The validity (and the quality) of modelling is crucially
dependent on:

(a)   the character of the data and

(b)   the quality of the learning set used in modelling.

In chemistry, various representations of chemical objects
were applied for different learning sets. Very common is
the representation of the objects by physico-chemical
data; the compounds can be represented, e.g., by their
spectral data (Sub-section 3.1.4), by electrochemical data
(Sub-section 3.1.5) and so on. Very important are the representations based on chemical structures (see Sub-
section 3.1.3). Especially useful are fundamental physico
-chemical parameters of the chemical elements as atomic
radius, electronegativity and others (Sub-section 3.1.2).

The use of the latter representation could be advantageous
with respect to availability as well accuracy of these
data and their high chemical information content. Unfortunately, the use of fundamental physico-chemical parameters is limited to the encoding of compounds of
limited complexity only. For such chemical systems the
pattern analysis seems to be one of the promising
approaches to modelling (Štrouf, 1980).

The first modelling of this type was tentatively described by Kowalski and Bender (1972a) for acidobasic
properties of representative oxides of chemical elements
characterized by six physico-chemical parameters. Recently, Štrouf, Kuchynka and Fusek (Štrouf et al., 1981a,
1981b; Kuchynka et al., 1981, 1982) have modelled catalytic activity of metals in heterogeneous catalytic
reactions. Modelling of catalytic activity of thirty
transition metals in hydrogenolysis of ethane can be made
on the basis of seven linearly independent variables
selected from the set of original fifteen variables using
SPHERE method (see Sub-section 2.1.3) (Štrouf et al.,
1981a). Analogously, the modelling of chemisorption of
hydrogen and carbon monoxide on metals has been carried
out because of their importance as initial steps in the
catalytic Fischer-Tropsch synthesis. In the former case
the correctness of recognition is high with nine linearly
independent variables selected from the original fifteen
physico-chemical variables (Kuchynka et al., 1981); the
reduced set of variables includes the variables related
to the Pauling equation for the adsorption heat of hydrogen and, moreover, six additional ones (Štrouf et al.,
1982). In the latter case, an analogous analysis of
chemisorption of carbon monoxide has been performed using
a set of eight linearly independent variables; furthermore, the dissociation of carbon monoxide chemisorbed on
transition metals has been modelled by heat of fusion,

molar heat capacity, molar electrical conductivity and electronegativity (Štrouf et al., 1981b; Fusek et al., 1983).

Pijpers and Vertogen (1982) have modelled the super-conductivity of the elements of the Periodic System; a theory powerful in predicting the superconducting behaviour from physico-chemical parameters is not available. Relevant variables in this problem are the electronic work function and the number of valence electrons as in the Miedema's model for alloy formation and, moreover, the specific heat, the heat of fusion, the heat of sublimation, the melting point and the atomic radius. The selection has been made from seventeen features the values of which are available in literature. Generally, this limitation is a very frequent case in pattern recognition modelling. Its result is that the set of variables cannot be commonly exhaustive. Nevertheless, in the case of modelling with basic physico-chemical parameters there is a high probability to find a set of sufficiently relevant parameters between the available ones.

It is worth mentioning that a combination of this physically meaningful representation with some other more formal encoding can also be useful. For example, in pattern recognition analysis of stability of complex hydrides $ABH_{4-n}D_n$ (where A is an alkali metal, B an element of the 3rd Main Group of the Periodic Table, H a hydride atom, D a ligand and n the number of these ligands), e.g. $LiAlH_4$, $NaBH_4$, $NaAlH_2(OCH_2CH_2OCH_3)_2$, the atoms A and B can be characterized by their physico-chemical parameters (Štrouf and Wold, 1977; Štrouf and Fusek, 1979; Wold and Štrouf, 1979, 1979a; Fusek and Štrouf, 1979).

The modelling by pattern recognition is restricted to those systems in which the objects are members of an inherently limited set (e.g. the elements of the Periodic

System) and are represented by physically meaningful
parameters (e.g. by atomic radius, electronegativity etc.)
on the basis of relatively high level of a priori knowledge of the system. The models for more complicated cases
(e.g. for spectral data) are not available and this is
the reason for the failure of some pattern recognition
analyses as recently discussed (Clerk and Székely, 1984).

In general, pattern recognition does not provide
theories but, in certain situations, can merely indicate
the importance of some variables and, therefore, it can
be useful for unbiased assessment of theories (Pijpers
and Vertogen, 1982).

CHAPTER 2

Methodology of Chemical Pattern Recognition

2.1 SUPERVISED CLASSIFICATION METHODS

In supervised methods, the classification is based on so called training (learning) sets of objects, i.e., on the sets of objects with known membership to some of the classes of the system under consideration. Throughout this book, the members of training sets are called prototypes. A set of as representative as possible prototypes $x^{(q)}$ should be gathered for Q classes of the system. The goal of supervised classification methods is to estimate parameters of some selected classification rule using these prototypes. The quality of the training sets determines principally the quality of the results of classification and, more importantly, whether these results have any validity at all.

2.1.1 DISCRIMINANT ANALYSIS

The most frequently used supervised classification methods are based on the concept of a discriminant function and they are termed discriminant analysis. The discriminant function $f(x)$ has the property that it partitions the measurement space into mutually exclusive regions (subspaces) characteristic for classes q. The discriminant

function is defined in such a way that for all objects
(points) $\underline{x}$ within the region describing the k-th class,
there exists a function $f_k(\underline{x})$ such that

$$f_k(\underline{x}) > f_l(\underline{x}) \qquad \text{for all } k \neq l$$

The surface separating regions for classes q_k and q_l is
given by the equation

$$f_k(\underline{x}) - f_l(\underline{x}) = 0$$

Hypothetically, $Q(Q-1)/2$ such separating surfaces can be
constructed in a system with Q classes. For a dichotomy
problem, i.e., the problem of dividing two classes only
($Q = 2$), a single discriminant function is constructed
as follows:

$$f(\underline{x}) = f_1(\underline{x}) - f_2(\underline{x})$$

The decision (classification) is made according to a very
simple rule: when $f(\underline{x})$ is positive, class q_1 is determined,
when $f(\underline{x})$ is negative then q_2 is determined.

  In discriminant analysis a variety of methods for constructing the discriminant function and evaluating its
coefficients may be used. Theoretically, many types of
mathematical functions can be used as discriminant functions. Nevertheless, in practical classification problems,
linear discriminant functions are most frequently
employed (Figs.3a and 3b). Among nonlinear functions such
quadratic ones are adopted that include terms of second
order. The quadratic discriminant functions can be transformed to a linear form by an appropriate preprocessing,
as described for example by Nilsson (1965, p.30).

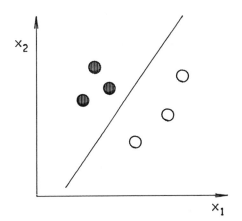

Fig.3a.
Separating line for two-
class problem in two-
dimensional space.

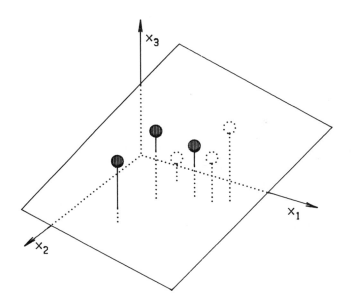

Fig.3b.   Separating plane for two-class problem in
           three-dimensional space.

Furthermore, there exist many methods for the adjustment of the coefficients (weights) of the selected discriminant function. In principle, these methods can be divided into nonparametric (distribution-free) methods and parametric ones.

Nonparametric discriminant analysis

Nonparametric discriminant analysis is employed in the cases when the distribution of the data is unknown, this situation being rather common in real problems. Therefore, the analysis cannot be made by estimating the values of the parameters of a distribution function. It is worth mentioning here that the nonparametric methods are based also on the estimation of parameters - but these parameters are not the parameters of a distribution function. The nonparametric approach is suboptimum in respect to the Bayesian parametric approach (Sub-section 2.1.1), but it permits a more versatile utilization than the latter, which embodies strict statistical assumptions. The nonparametric methods represent a well developed branch of supervised chemical pattern recognition (Jurs and Isenhour, 1975; Varmuza, 1980) which has attracted a considerable interest from chemists in the recent years. An assessment of the possibilities of these methods has been presented using different original methods (e.g., the Monte Carlo method, Stouch and Jurs, 1985, 1985a).

Learning machine

Very popular in chemical pattern recognition are linear discriminant functions with parameters (weights) trained iteratively by an adaptive method using negative feedback. This approach is commonly named learning machine (Nilsson, 1965). For practical purposes, the learning machine

algorithm involves a translation of the separating
(decision) hyperplane in such way that the hyperplane
passes through the origin of the N-dimensional measurement
space represented by the coordinate system formed by N
axes (N is the number of variables) (Fig.4).

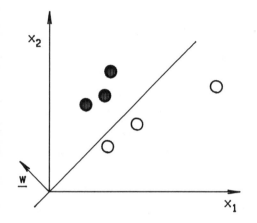

Fig.4.
Learning machine.
Vector $\underline{w}$ is the weight
vector of the machine.

This is achieved by adding a component w_{N+1} to the original weight vector and the component $x_{N+1} = 1$ to each
pattern.

The classification rule can be formulated as follows:
if the sign of the scalar product of the weight vector $\underline{w}$
and of the pattern $\underline{x}$

$$\underline{w} \cdot \underline{x} = w_1 x_1 + w_2 x_2 + \cdots + w_i x_i + \cdots + w_N x_N + w_{N+1} x_{N+1}$$

is positive, the pattern is assigned to the class q_1 and
if the product has the negative sign, the pattern is
classified into the class q_2

$\underline{w}.\underline{x} > 0,$ $\underline{x}$ is assigned into class q_1,

$\underline{w}.\underline{x} < 0,$ $\underline{x}$ is assigned into class q_2.

The training process (the estimation of the values of the components of the vector $\underline{w}$) consists of:

(a) selection of an initial weight vector $\underline{w}_1$, advantage-ously preselected by an appropriate simple method. For example, the vector $\underline{w}_1$ can be a vector defining the plane bisecting the distance between the centres $\underline{c}^{(q)}$ of the classes q_1 and q_2. The results of classi-fication can be significantly influenced by the selection of the initial vector in the training procedure.

(b) calculation of the scalar product $\underline{w}_1.\underline{x}_1$ using the values of x_i (measurements) for the first selected prototype $\underline{x}_1$. When the result is false (i.e., the calculated value has an opposite sign as expected) then the position of the hyperplane must be changed. The correction is made by changing the initial vector $\underline{w}_1$ by a correcting factor $c.\underline{x}_1$

$$\underline{w}_2 = \underline{w}_1 + c.\underline{x}_1$$

Several methods for correction of the hyperplane posi-tion have been described in chemical pattern recognition (see, e.g., Jurs and Isenhour, 1975, pp.13-15; Varmuza, 1980, pp.33-35). Most frequently used is the reflection of the plane across the incorrectly classified object

$$\underline{w}_2.\underline{x}_1 = {}^-\underline{w}_1.\underline{x}_1$$

and

$$(\underline{w}_1 + c\underline{x}_1)\cdot\underline{x}_1 = -\underline{w}_1\cdot\underline{x}_1$$

The correction coefficient c for the reflection is finally calculated according to the relation

$$c = -2(\underline{w}_1\cdot\underline{x}_1)/\underline{x}_1\cdot\underline{x}_1$$

The corrections are performed until all prototypes are subsequently employed. Such a training process suffers from the dependence of the results on the order in which the prototypes are included in the training.

Two principal situations can be found in the evaluation of the training results:

(a)   The system under study is composed of classes which are linearly separable; this means that all proto- types can be correctly assigned to their "own" class using the linear discriminant function and,

(b)   on the other hand, the system is composed of classes which cannot be separated by a linear hyperplane. Unfortunately, this linear inseparability cannot be known a priori. It is recognized only during iter- ative estimating of the weights when the training process does not converge (i.e., a completely cor- rect recognition of prototypes cannot be achieved even after multistep iteration).

In spite of this fact, the classification using learning machine approach is frequently used because of the ex- tremely simple calculation of the class membership for a new object; only the multiplication of the data x_i of the pattern $\underline{x}$ by the corresponding components w of the trained weight vector $\underline{w}$ and summation of the obtained products are necessary for the classification according to the above classification rule.

The above mentioned non-generality is, naturally, a serious inherent disadvantage in this method; the linear learning machine is an efficient classification tool only for linearly separable systems. Furthermore, this conceptually simple method gives only qualitative classification results without using additional valuable information contained in the system, which is feasible by adopting more sophisticated pattern recognition methods. For example, an outlier (an object far from every class of the system) cannot be detected and may dramatically adversely affect the training process. Moreover, for the training to be reliable, the ratio of the number of prototypes to the number of features must be high (at least 3).

Some attempts have been made to preserve the extremely simple linear form of discriminant function even for those systems in which the classes are not completely linearly separable. The conceptually simplest approach is to implant a zone along the decision hyperplane (Fig.5),

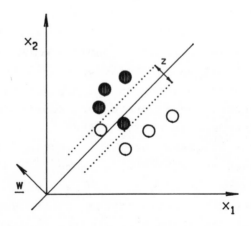

Fig.5. Learning machine with "dead" zone z.

the width of which is trained to be optimum (see Varmuza, 1980, pp.39-41). Naturally, the objects lying in this "dead" zone cannot be classified.

Simplex optimization of a linear discriminant function

The simplex optimization of the weights in linear learning machines was described earlier (Ritter et al., 1975; Brunner et al., 1976; Lam T.F. et al., 1976; Kaberline and Wilkins, 1978). Along with the advantage of the applicability for linearly inseparable classes, the possibility to optimize more complex discriminant functions should be mentioned. Some methodological improvements of simplex optimization method have been recently published (Brissey et al., 1979; Ryan et al., 1980; Štrouf and Fusek, 1981; Åberg and Gustavsson, 1982; Gustavsson and Sundkvist, 1985). The improved simplex of Štrouf and Fusek has been tested in the search for mathematical representation of classes structures in the SPHERE method (Sub-section 2.1.3).

Optimization of a linear discriminant function by least squares method

Another approach used for the optimization of linear discriminant functions for linearly inseparable classes is the linear regression based on the minimization of least squares of errors obtained in the comparison of calculated values for scalar products with those expected for the given set of prototypes. This method eliminates some constraints of the learning machine and is applicable even for overlapping clusters, in multicategory problems and in the problems in which the classification property of a continuous character should be predicted.

Mathematical treatments of this method and its variations
are given in Varmuza (1980, pp.43-48).

Recently, Moriguchi et al. (1980, 1981, 1981a) have
developed a new discriminant method for systems with
ordered classes using an Adaptive Least-Squares (ALS)
technique. The ALS method makes decisions for multicategory classification by a single discriminant function
which is formulated by a feedback adaptation procedure
in a simple linear form. The values of the weight vector
components are determined by least-squares adaptation.

Analog feedback linear learning machine

Commonly used forms of the linear learning machine (see
above) are digital feedback linear learning machines
which are trained by the feedback of a digital output of
their decision part (threshold element). On the other
hand, the Analog feedback Linear Learning Machine (ALLM)
(Ichise et al., 1980a) is trained by the feedback of an
analog output of its adaptation part. The ALLM can be
schematized as follows (Fig.6 on page 23). The ALLM is
composed of an input part and an adaptation part; as input,
a pattern $\underline{x}$ is used. In the adaptation part the pattern is
weighed by a weight vector $\underline{w}$ and the linear discriminant
function $f(\underline{x})$ is then calculated. The vector $\underline{w}$ is trained
so as to minimize the difference $\varepsilon(t)$ and the desired
value d

$$\varepsilon(t) = d - \sum_{i=1}^{N+1} w_i(t-1)x_i(t)$$

where $w_i(t-1)$ and $x_i(t)$ represent the i-th element of
$\underline{w}(t-1)$ and $\underline{x}(t)$, respectively. In the ALLM, $\underline{w}(t)$ is
changed according to

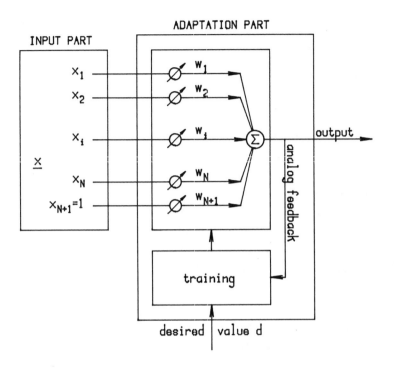

Fig.6.   Scheme of ALLM with input and adaptation parts.

$$\underline{w}(t) = \underline{w}(t-1) + m\epsilon(t)\underline{x}(t)$$

The principle of the analog feedback method is the
following inequality for the constant m:

$$0 < m < 2 \sum_{i=1}^{N+1} x_i^2(t)$$

A set of prototypes $\underline{x}(t)$ is presented to the ALLM cycli-
cally. The weight vector $\underline{w}$ converges to a vector which
minimizes the error $\epsilon(t)$. The simple ALLM algorithm is
suitable for quantitative problems and it has been recently employed for electrochemical purposes (Ichise et al.,
1980a, 1982).

Fisher discriminant analysis

In Fisher linear discriminant analysis, for a given data
set the direction in the measurement space is searched in
such a way that objects (points) projected onto a line in
this direction will be maximally discriminated in respect
to the two classes of the system. The criterion for discrimination is the Fisher ratio (Sub-section 2.3.2). This
ratio for the variables is maximized by an appropriate
orthogonal linear transformation, e.g., by that described
by Foley and Sammon (1975).
   Rasmussen et al. (1979a) have recently used a two-level
approach to display multivariate data for classification
purposes; they have determined two mutually orthogonal
discriminants that maximize the Fisher ratio. These two
discriminants are suitable as axes in a two-dimensional
graph (see Section 2.4).

Double stage principal component analysis

Hoogerbrugge et al. (1983) have recently described discriminant analysis by a double stage principal component
analysis (see Sub-section 2.3.1), which is suitable for
those cases when the number of variables is too large
compared with the number of objects. In the first stage,
the authors reduce the number of variables by principal
component analysis and afterwards they perform discri-

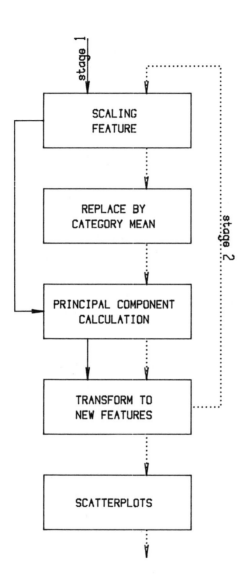

Fig.7. Scheme of double stage principal component
analysis.

minant analysis as a second principal component analysis
(Fig.7).

Multivariate and discriminant analysis

The program of Ahrens and Läuter (1974) called Multivariate and Discriminant Analysis (MVDA) consists of two
parts: the first one selects features (Section 2.3) and
the second one is concerned with calculating so-called
nonelementary discriminant functions as a linear combination of such features so that they possess the maximum separating power (Läuter and Hampicke, 1973; Dove et al., 1979;
Danzer et al., 1984).

Recently, Wold and Dunn (1983) have discussed the
condition:

the number of parameters $\ll$ number of degrees of freedom

$$P \ll DOF$$

as well as the related level of triviality (LOT):

$$P = DOF$$

for different multivariate data analysis methods including
linear discriminant analysis. This analysis is a regression
-like method having similar mathematical properties and,
therefore, the above condition $P \ll DOF$ yields $N + 1 \ll M$
and the level of triviality LOT is expressed by the relation $N = M - 1$ where N is the number of variables
(dimensionality) and M is the number of objects available
for estimation of the parameters (prototypes). The same
relation is also valid for the linear learning machine,
as an example of linear discriminant analysis methods.

Whalen-Pedersen and Jurs (1979) have derived the relation between the ratio M/N and the probability of dichotomization as a function of the distribution of the training set population. The relation was tested for a linear classifier using Gaussian and uniform distributed data.

A new representation concept (named the teaching space approach) has been introduced by means of a theorem, which makes transparent the essential factors of the linear pattern classifier training (Božinovski, 1985).

Recently, numerous publications describing the applications of linear learning machines in chemistry for various purposes have appeared and some of them are given here for illustration; they have been used, e.g., to:

- develop a measure of confidence of the results for classification of mass spectra (Richards and Griffiths, 1979),
- directly compare the results for six sets of real data with those of other four pattern recognition methods (Sjöström and Kowalski, 1979),
- predict the origin of olive oil (Forina and Tiscornia, 1982a),
- find features of chemical elements relevant for superconductivity behaviour (Pijpers and Vertogen, 1982),
- investigate the relation between structure and taste quality (Takahashi et al., 1982),
- estimate the concentrations of one strong and two weak bases from computer-generated potentiometric curves of their mixtures (Bos, 1979),
- analyse qualitatively non-linear static characteristics of the electrode process (Ichise et al., 1980),
- determine simultaneously the linear and non-linear characteristics of the electrode process (Ichise et al., 1982),

- determine metal ions concentration from staircase polarographic data (Ichise et al., 1980a),
- interpret mass spectra of nucleosides (Maclagan and Mitchell, 1980),
- classify compounds on-line on the basis of their gas chromatography-infrared data (Hohne et al., 1981),
- classify four oil types on the basis of their gas chromatographic data (Clark and Jurs, 1979),
- differentiate the aroma quality of soy sauce from gas chromatograms (Aishima et al., 1979) and
- classify monofunctional compounds using gas chromatographic data (Huber and Reich, 1980).

Furthermore, non-parametric discriminant functions have been also applied for:

- interpretation of analytical trace element data (Arunachalam et al., 1983),
- characterization of geochemical differences on the basis of analytical element data (Watterson et al., 1983),
- classification of crude oils from different lithostratigraphic situations on the basis of their trace element composition (Hitchon and Filby, 1984),
- prediction of olive oil origin from its fatty acid content (Forina and Tiscornia, 1982a),
- formulation of satisfactory quantitative structure–activity relationship (QSAR) (Chen B.K. et al., 1979),
- prediction of carcinogenic activity of N-nitroso compounds (Chou and Jurs, 1979),
- categorization of o-toluenesulfonyl thioureas and ureas into three glycemic activity classes (Dove et al., 1979),
- indication of therapeutic types of chemical compounds (Henry and Block, 1979, 1980),
- discrimination of hypotensive derivatives of guanidine (Moriguchi et al., 1980),
- studies on structure–activity relationships of anti-ulcerous and antiinflammatory drugs (Ogino et al., 1980),

- studies on structure-activity relationships for different biological activities (Moriguchi et al., 1981),
- analysis of structure-antitumour activity of withaferin analogues (Moriguchi and Komatsu, 1981a),
- studies on structure-carcinogenicity of different types of chemical carcinogens (Jurs et al., 1979), aromatic amines (Yuta and Jurs, 1981), N-nitroso compounds (Rose and Jurs, 1982) and polycyclic aromatic hydrocarbons (Miyashita et al., 1981),
- prediction of electrical conductivity of alloys and the activity of polymetallic catalysts (Belozerskikh and Dobrotvorskii, 1980),
- interpretation of partial chemical composition of complex biological samples on the basis of their pyrolysis-mass spectral data (Windig et al., 1983),
- characterization of samples from cultures of Bacteroides gingivalis using pyrolysis-mass spectral data (Boon et al., 1984),
- discrimination of several cigarette types on the basis of gas-chromatographic data for cigarette smoke (Parrish et al., 1981, 1983),
- extraction, reduction and ranking of gas-chromatographic data before using various statistical and graphical methods of pattern recognition packages (Hsu et al., 1982) and
- prediction of battery lifetime on the basis of initial cycling data (Perone and Spindler, 1984).

Parametric discriminant analysis

In those cases when the probability density of data belongs to the family of Gaussian (normal) distribution functions and, moreover, when the features are statistically independent, a parametric classification method can be used

(Varmuza, 1980, pp.78-82). When these two conditions are fulfilled (commonly are only assumed) the use of the Bayes classifier as an optimum one should be recommended.

In this approach, the discriminant function for a binary classification (dichotomy) is defined by the following equation:

$$f(\underline{x}) = \left[p(1)L(1)p(\underline{x}|1)/p(2)L(2)p(\underline{x}|2) \right] - 1$$

in which $p(q)$, $q = 1$ and 2, respectively, is a priori probability of class q (calculated as the ratio of the number of the respective prototypes M_q to the total number of prototypes M, M_q/M), $L(q)$ is the loss associated with misclassification of the prototype of class q and $p(\underline{x}|q)$ is the probability density of class q as a function of variables $x_1, x_2, \ldots, x_i, \ldots, x_N$.

The decision (classification) rule is the same as that for nonparametric discriminant functions, i.e., when the calculated value of $f(\underline{x})$ is positive the object (pattern) $\underline{x}$ is assigned to class 1, when it is negative then the pattern is classified into class 2.

For multicategory systems with Q classes the conditional probabilities for each class q are calculated according to a Bayes-like relation

$$p(\underline{x}|q) = p(q)L(q)p(\underline{x}|q)/ \sum_{q=1}^{Q} p(q)L(q)p(\underline{x}|q)$$

The object $\underline{x}$ is classified into the class with the largest value of the conditional probability. When the classes have equal a priori probabilities (they are uniformly populated) and the loss functions are symmetrical, the classification can be performed using only conditional probabilities $p(\underline{x}|q)$ and the approach is called the maximum likelihood classifier.

The most serious problem in parametric classification methods is therefore the determination of the conditional probability densities $p(\underline{x}|q)$ for Q classes (Andrews, 1972, pp.113-118)

$$p(\underline{x}|q) = \left[(2\pi)^{N/2}|\underline{\Phi}_q|^{1/2}\right]^{-1} \exp\left\{-1/2\left[(\underline{x}-\underline{\mu}_q)^T \underline{\Phi}_q^{-1}(\underline{x}-\underline{\mu}_q)\right]\right\}$$

where $\underline{\Phi}_q$ is the determinant of matrix $\underline{\Phi}_q$. For complete specification of the density function one needs only to calculate the expectations $\underline{\mu}_q$ and the covariance matrices $\underline{\Phi}_q$. These parameters can be estimated from prototypes as $\underline{m}_q$ and $\underline{U}_q$, respectively:

$$\underline{m}_q = (1/M_q)\sum_{k=1}^{M_q} \underline{x}_k^{(q)}$$

$$\underline{U}_q = (1/M_q)\sum_{k=1}^{M_q} (\underline{x}_k^{(q)} - \underline{m}_q)^T(\underline{x}_k^{(q)} - \underline{m}_q)$$

Under the validity of the feature independence condition, the conditional probability density can be estimated from the probability densities of the individual features

$$p(\underline{x}|q) = \prod_{i=1}^{N} p(x_i|q)$$

where N is the number of the features and $p(x_i|q)$ is the conditional probability density for a feature x_i in class q, which can be determined from the histogram (Fig.8).

32

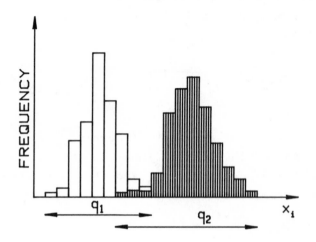

Fig.8. Histogram for the i-th variable in classes q_1 and q_2, respectively.

Under the conditions required for the applicability of maximum likelihood classifier the discriminant function can be calculated from the following relation

$$f(\underline{x}) = \sum_{i=1}^{N} \lg \, p(x_i|1)/p(x_i|2)$$

and the decision rule for discriminant analysis is used.

Above, the common density estimation using prototypes has been outlined. Nevertheless, the estimation of class-dependent probability densities $p(\underline{x}|q)$ can be carried out directly from prototypes. This approach is based on the methodology of the K-nearest neighbour method and the potential function method (see Sub-section 2.1.2).

Therefore, it suffers from high requirements on computation. On the other hand, no a priori knowledge of the density function is necessary.

Recently, Habbema (1983) has discussed sixteen extensions and adaptations of the standard model for probabilistic supervised pattern recognition divided into five main topics: the classification rule, measurement of performance, feature vector, training set and the actual pattern class of an object. Many of the extensions and adaptations are easy to incorporate in any probabilistic supervised pattern recognition approach.

Wold and Dunn (1983) have discussed the applicability of the Bayesian method. In the simplest case the probability density function involves only the mean and the standard deviation of each variable. This involves $2N$ parameters per class (N is the number of variables) estimated from $N \times M_q$ data (M_q is the number of prototypes in class q). For non-orthogonal variables, the covariances between the variables should be added and the number of parameters then increases to $2N + N(N-1)/2$ per class. Thus the level of triviality is for the simplest case with independent variables $M_q = 2$ and for the more realistic case with dependent variables $M_q = 2 + (N-1)/2$.

Ogino et al. (1980) have introduced "admissible" discriminant analysis (Anderson and Bahadur, 1962) in chemistry. The method is useful in the case when the condition for usual statistical discriminant analysis, i.e., the equality of the covariance matrices for normally distributed variables, is not fulfilled. In the procedure described, the model of equal covariances is not the prerequisite for the analysis. As a primary criterion for selecting the best combination of variables in the discriminant functions, the minimum number of misclassified compounds was used. The "admissible" procedure searches for the "most reasonable" discrimination functions, i.e.,

the "best" covariance matrix common to two groups estimated in terms of a linear combination of unequal covariance matrices of two classes by an iterative procedure.

Coomans et al. (1981c) have described a combination of
"display" Statistical Linear Discriminant Analysis (SLDA)
with the ALLOC method (see Sub-section 2.1.2). The
"display" SLDA means here linear discriminant functions
that are linear combinations of the original variables
transforming measurement space to the factor space, which
can be of low dimensionality $1 \leqq D \leqq 3$. The combination
permits adjustment of the suboptimum decision boundaries
which are obtained by classification with the usual SLDA,
even in the cases when the assumptions about the data made
in the SLDA are not fulfilled. The procedure consists of
the following steps: (a) preprocessing of the original
data by linear combination of the features with discriminant weights and plotting of the objects in a diagram,
and (b) calculation of the a posteriori probabilities
of class membership for the objects represented in the
diagram. This is done by the ALLOC classification of the
discriminant scores of the objects (Hermans and Habbema,
1976).

Sjöström and Kowalski (1979) have discussed the fact
that the class distributions estimated from the training
set lead to an over-optimistic recognition result.
Moreover, when only a few prototypes are available for
the estimation of the class distribution, the risk of
incorrect prediction increases. Therefore, for chemical
applications where the class distributions are rarely
known and small data sets are frequent, Bayes classifier
should be used with caution.

Statistical Bayesian discriminant analysis has been
recently used in various fields of chemistry for very
different purposes, e.g., for:

- comparison with other four pattern recognition methods using six sets of real data (Sjöström and Kowalski, 1979),
- modelling multivariate quantitative structure-activity relationships (QSAR) and other scientific models (Wold and Dunn, 1983),
- differentiation between thyroid functional states (in combination with the ALLOC) (Coomans et al., 1981c),
- classification of oils into four oil types (Clark and Jurs, 1979),
- prediction of Italian olive oil origin by their fatty acid content (Forina and Tiscornia, 1982a),
- prediction of elemental composition and substitution types of nucleosides from their mass spectra (Maclagan and Mitchell, 1980),
- classification of monosubstituted phenyl rings characterized by combined infrared and Raman spectral data (Tsao and Switzer, 1982; Tsao, 1982b) and
- analysis of local geometrical features of ion solvation by treatment of the results of Monte Carlo simulations (Marchese and Beveridge, 1984).

2.1.2 K-NEAREST NEIGHBOUR RULES AND POTENTIAL METHODS

K-Nearest neighbour rules

A substantially different concept in comparison with discriminant analysis (Sub-section 2.1.1) underlies the classification methods called K-Nearest Neighbour (KNN) methods. They search primarily for similarity within classes, whereas discriminant methods search for dissimilarity between classes. KNN methods are based, in their principal version, on:

(a) the determination of all distances between the
     classified object and the prototypes,
(b) the ranking of these distances,
(c) the estimation of the number of nearest neighbours
     in prototypes of the classes q_k ($k = 1,2,...,k,...,Q$),
(d) the classification of the object into the class q_k
     with the maximum number of the nearest neighbours
     belonging to the class q_k.

The number of nearest neighbours K is commonly selected
from odd integers, for unique voting to be possible.
In Fig.9 an example for K = 3 is given; two votes are for
class q_1 and only single one for class q_2. Therefore the
object $\underline{x}$ is classified as a member of the class q_1.

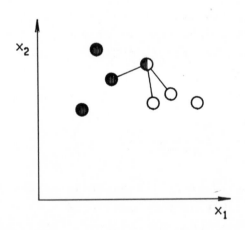

Fig.9.   KNN rule for K = 3. The classified object ◑ has
         two nearest neighbours from the class q_1 (○)
         and one from class q_2 (●).

The value of K has an effect on the result of classification as can be illustrated in Fig.10, where the class
membership varies with different K.

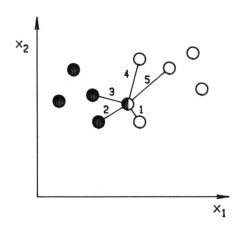

Fig.10. Effect of the value of K in KNN rule on the
classification. Object ◐ belongs to class
q_1 (○) for K = 1 and 5, and to class q_2 (●)
for K = 3.

Nevertheless, the KNN method is considered as a standard
method because it works with "natural" data. Unfortunately,
high requirements on computation are inherent in the fundamental version of this method. Each classification of
a new object needs the recalculation of all distances and
new searching for K nearest neighbours.

This disadvantage is eliminated when one calculates a
representative point for the classes (usually the centre)
at first and then works with these representatives instead
of whole sets of the respective prototypes. The decision
on the classification of the object x is then made by
comparison of this object with all Q calculated centre

points c_k. It is evident that such a representation can
be oversimplified and, therefore, that it is valid only
in the cases of appropriately structured systems. In other
situations (not a priori known), this representation
cannot be used. For instance, from Fig.11 it is clear that
object x is incorrectly classified into class q_1 because
$d(x, c_1) < d(x, c_2)$, although it is evidently situated in
the region of class q_2.

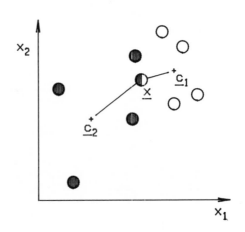

Fig.11.   Incorrect classification of object ◐ using
          distance measurement from centres of clusters.

   In spite of this deficiency, condensed KNN rules are
used because they reduce drastically the tremendous
amount of storage as well as computation requirements of
the original non-reduced version.
   Wold and Dunn (1983) have recently discussed the level
of triviality, LOT, for the KNN method. This method does
not work with any adjustable parameters and, therefore,

the LOT corresponds to one object per class. For instance, for the dichotomy problem the LOT is M = 2. If the scaling of variables based on class separation is made (for example Fisher weighting, see Sub-section 2.3.2) then N scaling parameters are calculated, and the LOT is N = M.

Coomans and Massart (1982) have described a so called alternative KNN method based on alternative voting. This method eliminates one of the shortcomings of common KNN method using a majority vote procedure which is not efficient in the case of systems consisting of overlapping classes considerably different in size as, for instance, those in Fig.12.

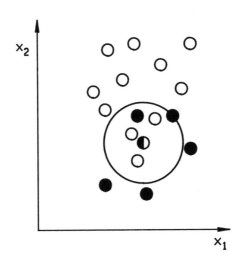

Fig.12. Alternative KNN method for the case of over-lapping classes of different sizes.

For dichotomy, the alternative vote might be formulated as follows: classify the object $\underline{x}$ in class q_1 if for a given K at least k (k = 1,2,...,K) neighbours belong to q_1,

40

otherwise classify the tested object in q_2. Naturally, one
of the possible values for k corresponds to the majority
vote rule, e.g., for K = 3,5,7,... these values are k =
2,3,4,... and so on. The optimum value of k is determined
using the classification rate obtained for different alternative voting rules (different k) as determined by the
leave-one-out evaluation method (Sub-section 2.1.5). The
use of the alternative voting rule can be extended to a
probabilistic KNN analysis.

Coomans and Massart (1982a) have applied in chemistry
the KNN method of Loftsgaarden and Quesenberry (1965) to
a probabilistic classification based on direct density
estimation. A given number $K^{(l)}$ of nearest neighbours are
selected for each training set q_l. The distance to the
farthest selected neighbour for the given class is a
measure of the likelihood of the object to that class.
The decision (classification) rule can be then defined as
follows:

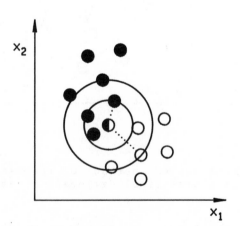

Fig.13. The KNN method of Loftsgaarden and Quesenberry.

classify the object to the class of maximum likelihood;
in other words, assign the point to the class for which
the distance (commonly Euclidean distance) is the smallest
one (Fig.13). The distance is represented by the radius r
of a hypersphere centered in $\underline{x}$. The volume of the hypersphere $V(\underline{x}/q_l)$ is inversely related to the probability
density $p(\underline{x}|q_l)$ (Loftsgaarden and Quesenberry, 1965)

$$p(\underline{x}|q_l) = (K^{(l)} -1)/n\left[V(\underline{x}/q_l)\right]$$

Coomans and Massart (1982a) calculated the volume of the
hypersphere according to

$$V(\underline{x}/q_l) = 2\pi^{N/2}r^N/N\Gamma(N/2)$$

where Γ is the gamma-function

$$\Gamma(x) = \int_0^{+\infty} e^{-t}t^{x-1}dt \qquad (x>0)$$

Furthermore, Coomans and Massart (1982b) have proposed
a new Condensed Nearest Neighbour (CNN) rule for non-
hierarchical clustering based on an operational research
model (Massart et al., 1980). The condensation is carried
out by the determination of p centrotypes. The method may
be used for probabilistic decision making, evaluated by
probabilistic measures for the performance and reliability.
The CNN method can reduce the data storage significantly
and, therefore, it might be a suitable non-parametric
classification procedure even for small computers.

Byers and Perone (1980) have proposed a new method for

feature weighting which has been optimized for the KNN
classification rule. This weighting is at least as good
as other weighting methods (see Sub-section 2.3.2) and it
can be advantageously combined with the selection of
features (Section 2.3).

Evidently due to a relatively simple conceptual basis,
several applications of the KNN rules have been recently
performed and some of them are presented here for illustration; for example, the KNN method has been used for:
- prediction of Italian olive oils origin using their
  fatty acid content (Forina and Tiscornia, 1982a),
- classification of mineral waters from different regions
  (Scarminio et al., 1982),
- distinguishing of coins from various mints of the Roman
  Empire on the basis of their elemental composition
  (Borszéki et al., 1983),
- on-line identification of dipeptide derivatives using
  their data from gas chromatography-mass spectrometry
  (Ziemer et al., 1979),
- classification of normal and virus-infected serums
  characterized by gas chromatographic data (Zlatkis et
  al., 1979),
- prediction of elemental composition and substitution
  types of nucleosides (Maclagan and Mitchel, 1980),
- evaluation of complex mixtures of non-volatile compounds
  characterized by their mass spectrometric data
  (van der Greef et al., 1983),
- distinguishing of functional groups and the presence of
  specific elements in vapour phase infrared spectra
  (Delaney et al., 1979),
- on-line classification of peak multiplicity in stationary electrode polarography (DePalma and Perone, 1979),
- indication of therapeutic type of compounds represented
  by their structure (Henry and Block, 1979),
- prediction of battery lifetime (Perone and Spindler,

1984) and
- identification of the nearest neighbour in chemical
  structure files for a more economical search (Willett,
  1983).

Potential methods

Potential methods are conceptually related to KNN methods.
Instead of a distance, an appropriate function is used
which has its maximum value centered in the prototype $\underline{x}^{(q)}$
and decreases symmetrically in all directions. Desirable
characteristics of such potential functions $g(\underline{x}, \underline{x}^{(q)})$ can
be formulated as follows (Andrews, 1972, p.88):
potential function should be
(a)   maximized for $\underline{x} = \underline{x}^{(q)}$,
(b)   approximately zero for $\underline{x}$ sufficiently far from $\underline{x}^{(q)}$,
(c)   smooth (continuous) and should decrease approximately
      monotonically with distance $d(\underline{x}, \underline{x}^{(q)})$ and
(d)   finally, if $g(\underline{x}_1, \underline{x}^{(q)}) = g(\underline{x}_2, \underline{x}^{(q)})$, then patterns
      $\underline{x}_1$ and $\underline{x}_2$ should have approximately the same "degree
      of similarity" to $\underline{x}^{(q)}$.
The form of this potential function is the matter of
selection. A variety of potential functions can be used;
the most commonly employed are a discrete triangular
function (Fig.14a) or a continuous Gaussian function
(Fig.14b) (see Coomans et al., 1981).
The cumulative potential function for a class q in the
point $\underline{x}$ is calculated by summation of the values of potential functions produced by the prototypes of the class q
in this place. For a dimensionality higher than two, one
obtains a potential hypersurface. Alternatively, the mean
potential function is used in the ALLOC method for density
estimation (see below).

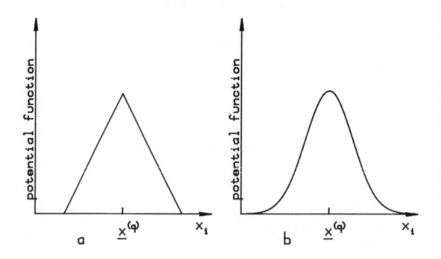

Fig.14. (a) Triangular and (b) Gaussian potential
functions.

In the point where a new object $\underline{x}$ is situated, values
of the cumulative or mean functions for the individual
classes are then compared. The decision on the classification of $\underline{x}$ is made in favour of the class q for which the
maximum value of density was found.

For classification purposes it is desirable to have
potential functions that possess a positive value within
the region of the class q, whereas, between the classes,
the value equals zero. This situation can be achieved by
smoothing the potential surface using appropriate coefficients; the selection of this smoothing parameter is
(along with choosing the type of the potential function)
another adjustable point in the classification method
based on potential functions.

Recent developments of potential methods for chemical purposes have been carried out mainly by the Massart's group. This group has introduced a new potential method in chemistry (Coomans et al., 1981), using the ALLOC package of Hermans and Habbema (1976). In this method, the Gaussian potential function is preferred to the other types of potential functions in constructing the mean potential function, which is used instead of the cumulative one; the mean function permits probabilistic classification of the Bayes type with probability density estimated in terms of potential functions of prototypes. In ALLOC, the smoothing parameter for each class q is calculated on the statistical basis using the sample variance for variable x_i obtained from the prototypes $\underline{x}^{(q)}$ of class q.

Theoretically it is impossible to estimate the probability density $p(\underline{x}|q)$ in a point $\underline{x}$. Therefore, the density is referred to an infinitesimally small hypercube with sides $d\underline{x}$. The probability density can be calculated from the relation

$$p(\underline{x}|q) = f(\underline{x}|q)d\underline{x}$$

in which

$$f(\underline{x}|q) = (1/M_q) \sum_{k=1}^{M_q} g(\underline{x}, \underline{x}_k^{(q)})$$

where M_q is the number of prototypes $\underline{x}_k^{(q)}$ $(k = 1, 2, \ldots, M_q)$ in class q.

As mentioned above, $g(\underline{x}, \underline{x}^{(q)})$ is a normal (Gaussian) function in the ALLOC. The general equation of an N-variate Gaussian potential function for prototypes $\underline{x}_k^{(q)}$ is

$$g(\underline{x}, \underline{x}_k^{(q)}) =$$
$$= \left[(2\pi)^{N/2} |\textstyle\sum_q|^{1/2} \right]^{-1} \exp\left\{ -1/2 \left[(\underline{x}-\underline{x}_k^{(q)})^T \textstyle\sum_q^{-1} (\underline{x}-\underline{x}_k^{(q)}) \right] \right\}$$

where $|\sum_q|$ is the determinant of matrix $\sum_q$. This matrix is a diagonal matrix, if the condition of non-correlated variables is fulfilled, and then it represents a set of adjustable smoothing parameters. In ALLOC, the smoothing parameter matrix $\sum_q$ for each class q is given by

$$\sum_q = c_q^2 \cdot \begin{bmatrix} s_{q,1}^2 & \cdots & 0 & \cdots & 0 \\ & \ddots & & & \vdots \\ 0 & \cdots & s_{q,i}^2 & \cdots & 0 \\ \vdots & & & \ddots & \vdots \\ 0 & \cdots & 0 & \cdots & s_{q,N}^2 \end{bmatrix}$$

where $s_{q,i}^2$ is the prototype variance obtained for the training set of class q for variable x_i ($i = 1,2,\ldots,N$). One adjustable smoothing parameter c_q^2 for each training set is estimated by a jack-knife maximum likelihood method (Toussaint, 1974).

The ALLOC method can estimate also the probability density for an action-oriented classification procedure (Coomans et al., 1982c) on the basis of the Bayes equation (see Sub-section 2.1.1). The action-oriented decision making is encountered mostly in clinical analytical chemistry, but also in some non-clinical problems such as the decisions associated with food quality control. The classification with ALLOC can be carried out in two different ways (Coomans et al., 1982c). First, the minimum probability of error classification rule is replaced by the

minimum overall risk rule, i.e., the boundary between two
classes is displaced so that it is closer to the class
with smaller misclassification risk. In the second way,
a region containing doubtful cases is defined, i.e., the
boundary is a zone instead of a line (compare the Sub-
section 2.1.1).

The ALLOC method belongs to the methods that include
direct estimation of the probability density. In this
sense it resembles KNN methods, but it differs from them
by taking the distance between $\underline{x}$ and the prototype $\underline{x}^{(q)}$
(neighbour) into consideration.

Recent applications are mostly related to the development of the ALLOC method (Coomans et al., 1981). For
testing purposes, the following problems have been studied:
- discrimination of the functional state of the thyroid
  gland as normal, hyper-active and hypo-active using
  clinical tests (Coomans et al., 1981, 1981b, 1981c),
- differentiation of pure milk from different species and
  mixtures (Coomans et al., 1981b).
Moreover, the "classical" Fisher's Iris data (Fisher,
(1936) have been used in testing ALLOC (Coomans et al.,
1981).

2.1.3   MATHEMATICAL MODELLING OF CLASSES

A recent approach in pattern recognition is represented
by the use of classification methods based on the assumption that the classes of the studied system can be described by a mathematical model (see also Section 1.2).
This approach has been recently studied by Svante Wold
in his SIMCA method (Wold, 1976; Wold and Sjöström, 1977)
and most recently in the SPHERE method by Štrouf and Fusek
(Štrouf and Fusek, 1979; Fusek and Štrouf, 1979). The use
of mathematical modelling provides remarkable advantages

and seems to be highly versatile. Generally, the models
provide a more fundamental insight into the regularities
in the data and, in this respect, they are superior to
other multivariate data processing.

Both the SIMCA and SPHERE methods belong to disjoint
methods, in which the classes are treated independently.
This makes possible to use pattern recognition also for
systems with one or several unstructured classes. In the
real world, such a situation is not rare and it arises
when one of the classification properties requires that
the parameter values lie in a limited range of possible
values, whereas the second property is related to widely
dispersed values of the parameters. This is very frequently found in the quality control and in the chemical
structure-biological activity problems (Dunn and Wold,
1980a) (see Sub-sections 3.1.3 and 3.2.3). For example,

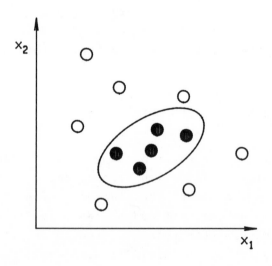

Fig.15. "Asymmetric" system.

in the latter case the activity of chemical compounds can
be characterized by a very limited range of descriptor
values whereas all other values can operate for inactive
compounds. This is illustrated by an artificial two-
dimensional example in Fig.15. The mathematical modelling
methods (e.g., SIMCA and SPHERE) are able to solve also
these cases (named by Wold "asymmetric" cases) by modelling only the structured class(es) of the system.

The SIMCA method

Soft Independent Modelling of Class Analogy (SIMCA) method
has been developed on a theoretical basis by Svante Wold
(Wold, 1974, 1976). Under some conditions (see below),
a linear model of principal components type can be in-
ferred as a similarity model (Wold and Sjöström, 1977).
A data matrix $\underline{X}^{(q)}$ for prototypes $\underline{x}_j^{(q)}$ of a class q with
elements (observed values of a variable) $x_{ij}^{(q)}$ can be
approximated arbitrarily closely (Wold, 1976) by the
relation

$$x_{ij}^{(q)} = \alpha_i^{(q)} + \sum_{a=1}^{A_q} \beta_{ia}^{(q)} \vartheta_{aj}^{(q)} + \epsilon_{ij}^{(q)}$$

where $\alpha_i^{(q)}$ is the mean of the variable x_i, $\beta_{ia}^{(q)}$ is the
loading of this variable, $\vartheta_{aj}^{(q)}$ is the prototype specific parameter and a = 1,2,...,A_q is the index denoting A_q
components of the model. The parameters $\alpha_i^{(q)}$, $\beta_{ia}^{(q)}$ and
$\vartheta_{aj}^{(q)}$ are estimated by the NIPALS method (see e.g. Wold
et al., 1983b) so that the residuals $\epsilon_{ij}^{(q)}$ are minimal.
   The model is valid if the following two assumptions are
fulfilled:

(a) The data must have a smooth character, i.e., the function generating these data must be several times differentiable so it can be expanded in the Taylor series. This assumption is generally fulfilled for measured data, whereas for discrete data (as are, e.g., structure descriptors) the results can be dubious (Wold and Sjöström, 1977).

(b) The objects within the class should be sufficiently similar so that the number of terms in the Taylor expansion can be limited and, consequently, the A_q in the model is reasonably small. In the SIMCA method, the number of components A_q is estimated by so called cross-validation procedure (Wold, 1978).

The SIMCA method allows the utilization of the information in multivariate data analysis regardless of the ratio between the number of variables and the number of prototypes (Wold et al., 1981). The level of triviality (LOT) for one-component SIMCA model ($A_q = 1$) is $2N + M_q - 1 = NM_q$. When the number of variables N is considerably higher than the number of prototypes M_q, i.e., $N \gg M_q$, LOT reduces to approximately $M_q = 2$. Thus, like the KNN method (but unlike the linear discriminant analysis including linear learning machine), the SIMCA method can be used in cases when the number of variables greatly exceeds the number of prototypes (Wold and Dunn, 1983).

The only restriction is that the number of components A_q must be smaller than both N and M. In fact, the classification stability of the SIMCA increases with the square root of N (for a given number of prototypes M). The practical minimum size is about five prototypes per class, but ten or more are preferrable since that provides opportunities for finding outliers and other detailed structure information on the class (Wold et al., 1983b). The inhomogeneities within the training sets can be revealed by the visualization using a principal components graph.

The NIPALS algorithm used in the adjustment of the model works also with an incomplete data matrix $\underline{X}$ when only less than 5-10% of the values are missing (Wold et al., 1983b).

Geometrically, the model can be represented by an A_q-dimensional hyperplane in the N-dimensional space with N variables. For $A_q = 1$ this corresponds to a straight line and for $A_q = 2$ to a plane. The representation for these situations in a three-dimensional space is given in Fig.16.

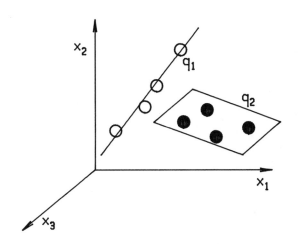

Fig.16. Disjoint principal component model (SIMCA). Class q_1 is described by a model with $A_q = 1$ and class q_2 by a model with $A_q = 2$.

The model includes a statistically determined confidence envelope, i.e., a space around the mathematical model defined by "typical" standard deviation estimated by the use of prototypes (Fig.17). The total residual standard deviation $RSD^{(q)}$ for the class q of the model can be calculated using formula (Wold, 1976; Wold and Sjöström, 1977):

52

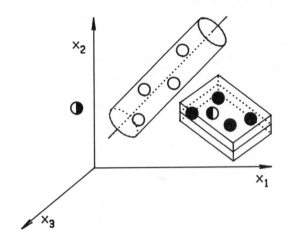

Fig. 17. The same as in   Fig. 16, but with   confidence
envelopes. The new point ◑ is inside   the
confidence limit of class q_2 and it is therefore
classified into this class. Another new point
◐ is outside   both classes and it is considered to be an outlier.

$$RSD^{(q)} = \left[\sum_{i=1}^{N}\sum_{j=1}^{M_q}(\epsilon_{ij}^{(q)})^2/(N - A_q)(M_q - A_q - 1)\right]^{1/2}$$

In the classification step, the object is fitted to the
class model using the multiple linear regression; each
residual standard deviation of object $\underline{p}$, $RSD_{\underline{p}}^{(q)}$, is
calculated from

$$RSD_{\underline{p}}^{(q)} = \left[\sum_{i=1}^{N}(\epsilon_{i,\underline{p}}^{(q)})^2/(N - A_q)\right]^{1/2}$$

The F-ratio of $RSD^{(q)2}$ and $RSD_p^{(q)2}$ is an indication of the goodness of fit (similarity) to the class q and can be used for classification purposes by choosing empirically a limiting optimum F-value.

Recently, Bink and van't Klooster (1983) have shown that the equivocation E, the expected value of the uncertainty after classification (Cleij and Dijkstra, 1979), can be adopted as an optimization criterion.

Another factor, which can be extracted from the data by means of the SIMCA polyalgorithm, is the distance between classes (Wold, 1976; Wold and Sjöström, 1977). When all M_r prototypes in class r are fitted to the model of class q and vice versa, the resulting RSDs can be used to obtain a measure of the separation of the classes - the class distance. The RSD of the prototypes in class r when fitted to class q is calculated from

$$RSD_r^{(q)} = \left[\sum_{i=1}^{N} \sum_{j=1}^{M_r} (\epsilon_{ij}^{(q)})^2 / (N - A_q).M_r \right]^{1/2}$$

Then, the symmetric distance between classes r and q is given by

$$d^{(r,q)} = \left[(RSD_r^{(q)2} + RSD_q^{(r)2})/(RSD^{(q)2} + RSD^{(r)2}) \right]^{1/2} - 1$$

The distance matrix gives the same picture as the eigenvector projection but in a quantitative way. In practice, distances smaller than 0.8-1.0 correspond to a small difference between classes and distances smaller than 0.5 to negligible class difference (Blomquist et al., 1979).

Principal component modelling is highly effective in all four levels of pattern recognition (Albano et al., 1978). The SIMCA method is used for the first two levels and

the MACUP (Modelling And Classification Using PLS) method
has been described for two-block principal component
modelling at levels 3 and 4 (see Wold et al., 1984).
The SIMCA method can be employed in the simple classification based on discrimination between a given set of
classes (level 1) with accounting for additional classes
including those with only single members - outliers
(level 2).

A remarkable recent development has been achieved predominantly for levels 3 and 4 (Wold et al., 1983a, 1984;
Martens et al., 1983). These levels are typical in numerous practical situations in which, in addition to the
classification made at levels one and two, the prediction
of values of dependent variables y (classification properties) is also needed. Partial Least Squares (PLS) regression on latent variables (Wold H., 1981) can be used
for the solution of such problems. At the beginning of
the solution, there stand two matrices (blocks) of quantitative variables: the first block of independent variables x_i (i = 1,2,...,N) is of the same kind as in situations at levels 1 or 2 and the second block consists of
dependent variables y_l (l = 1,2,...,L). The method is
based on the idea that both blocks can be described
simultaneously by a small number of the same factors
(latent variables).

On level three, we have a single dependent variable
(L =1) and the problem can be solved by the procedure
called PLS1 (Martens et al., 1983): The first factor
(latent variable) representing a linear combination of
all independent variables x_i, is estimated so that the
dependent variable y is predicted optimally in terms of
least squares residuals. Both x_i and y are projected onto
this factor. The second factor representing a linear combination of the residuals of x_i after the first projection, is then estimated. This factor is orthogonal to

the first one and it predicts the residual of x_i after the first projection in an optimal way. The x_i and y residuals are projected onto this second factor and so on.

At level four, the second block involves several independent variables y_L ($L > 1$); here, the PLS2 procedure has been successfully applied in a chemical application (Martens et al., 1983): the first factor is estimated iteratively so that all x_i are predicted optimally in terms of the least squares sum in residuals for all y_L. The variables x_i and y_L are then projected onto this factor. The second factor is estimated iteratively in next step. All the x_i and y_L residuals are then projected onto this second factor, which is orthogonal to the first one. The other factors are estimated in the same way.

The SIMCA method provides a possibility of scaling the variables either over all prototypes (classical scaling in multivariate analyses) or only over the prototypes belonging to the same individual class (separate scaling). It has been demonstrated (Derde et al., 1982a) that the latter scaling has a beneficial effect on the classification (compare also the Section 3.1).

There have been made many recent applications of the SIMCA method, especially in:

- classification of six sets of data for direct comparison of its performance with that of other four different pattern recognition methods (Sjöström and Kowalski, 1979),
- searching for basic regularities in the stability of unsubstituted complex hydrides ABH_4 (Wold and Štrouf, 1979) and monosubstituted complex hydrides ABH_3D (Wold and Štrouf, 1979a),
- modelling of quantitative structure-activity relationships in the cases where the structures are characterized by quantitative variables and the biological activity by measured values (Dunn and Wold, 1980, 1981;

Wold and Dunn, 1983),

- modelling of substituent effects in NMR spectra of
  4-substituted styrenes (Edlund and Wold, 1980),
- classification of fungal samples of unknown origin by
  analysis of data obtained from repetitive pyrolysis-
  gas chromatograms (Blomquist et al., 1979, 1979a),
- discrimination between normal brain tissue and tumours
  using gas chromatographic data (Jellum et al., 1981;
  Wold et al., 1981),
- selection of biochemical characteristics in the breeding
  for pest and disease resistance by analysis of gas
  chromatographic data (Lundgren et al., 1981),
- separation of botulinum-positive and negative fish
  samples by means of their chromatograms (Snygg et al.,
  1979),
- sensory evaluation of aroma in soy from chromatographic
  data (Aishima et al., 1979),
- classification of fruit bodies of some Ectomycorrhizal
  Suillus species using pyrolysis-gas chromatography data
  (Söderström et al., 1982),
- examination of geochemical data from chemical analysis
  of major and trace elements (Bisani et al., 1983),
- application in geochemistry and geology (Esbensen and
  Wold, 1983),
- study of environmental pollution by analysing gas
  chromatograms of samples of blue mussels (Kvalheim
  et al., 1983) and
- classification of chemical structure using autocorre-
  lated mass spectra (Wold and Christie, 1984b).

Furthermore, the SIMCA version of the principal component analysis has been employed in a multivariate analysis
of ionization constant data (Edward et al., 1981). Classification of organic compounds has been made by principal
component analysis of their infrared spectra (Bink and
van't Klooster, 1983). Hoogerbrugge and co-workers (1983)

have described a discriminant analysis by double stage
principal component analysis (see Sub-section 2.1.1).

Partial least-squares (PLS) path modelling with latent
variables has been used in modelling of various problems:
- It has been used in modelling the water flow downstream
  quality (Gerlach et al., 1979).
- A PLS algorithm has been employed for analysis of ^{13}C-
  NMR parameters-carcinogenic activity relation (Nordén
  et al., 1983).
- The PLS method has been used to estimate quantitative
  structure-activity relationships between multivariately
  described chemical structure and a battery of biological
  tests (Dunn et al., 1982).
- The PLS methods have been examined for geochemical
  data from elemental analysis (Bisani et al., 1983).
- A PLS data analysis has been used to predict the ^{13}C-
  NMR shifts (Johnels et al., 1983).
- The PLS method has been used to resolve fluorescence
  spectra of complex mixtures and to quantify components
  of the mixtures (Lindberg et al., 1983).
- The PLS method has been used for comparing two blocks
  of variables: sensory data from cauliflower (Martens
  et al., 1983).
- The PLS method can be used for multivariate calibration
  problems in analytical chemistry as an alternative to
  principal component analysis combined with multiple
  regression (Sjöström et al., 1983).
- The PLS technique of pattern recognition has been used
  for the detection of constituents in complex mixtures
  on the basis of gas chromatography (Dunn et al., 1984).

Recently, an extension of SIMCA (named CLASSY) has been
made by applying kernel density estimation, borrowed from
ALLOC, to the scores inside the class model subspace in
combination with a normal error distribution in the
remaining dimensions (Van der Voet and Doornbos, 1984a).

The performance of this new probabilistic classification method CLASSY has been compared with the performance of its predecessors SIMCA and ALLOC (van der Voet and Doornbos, 1984b).

The SPHERE method

Recently, a new disjoint pattern recognition method, which models each class of system as a hypersphere, has been developed (Štrouf and Fusek, 1979; Fusek and Štrouf, 1979) and it is called here SPHERE (Similar Patterns HEuristic REcognition). The classification part of the SPHERE method can be described by the following algorithm (Fusek and Štrouf, 1979):

(1) The data x_{ij} of each training set are autoscaled so that the mean be zero and the standard deviation be unity (see Section 3.1).

(2) The covariance matrix $\underline{U}_q$ is formed for the individual training set (see Sub-section 2.3.1).

(3) For each class q, the eigenvectors $\underline{e}_i$ and the eigen-values λ_i are calculated from the matrix $\underline{U}_q$ (see Sub-section 2.3.1).

(4) The autoscaled data of the prototypes x'_{ij} in the class are transformed by means of the matrix of $\underline{e}_i$ according to

$$w_{ij} = \sum_{r=1}^{N} x'_{rj} \cdot \underline{e}_{ir}$$

where $i, r = 1, 2, \ldots, N$ (N is the number of variables) and $j = 1, 2, \ldots, M_q$ (M_q is the number of prototypes).

(5) The eigenvalues $\lambda_i^{1/2}$ calculated in step (3) are ranked and only those giving the value of $\lambda_i^{1/2}$ higher than that estimated as $\lambda_{error}^{1/2}$ (see p.63) are considered in the following step.

(6) The data for prototypes in the remaining dimensions R with $\lambda_i^{1/2} > \lambda_{error}^{1/2}$ are normalized according to

$$z_{ij} = w_{ij} / \lambda_i^{1/2}$$

the class being then defined, in a geometrical approximation, as a hypersphere with the radius (standard deviation) equal unity (Fig.18).

(7) The mean of squared distances $D_q^{(c)}$ of all prototypes of the class q from the centre C in normalized hyperspace is estimated according to

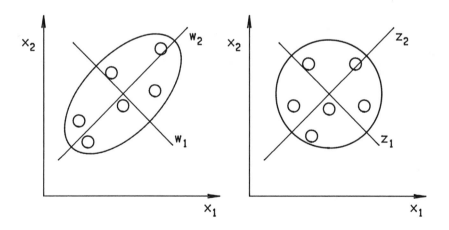

Fig.18. Normalization in the SPHERE method.

$$D_q^{(c)} = \left[\sum_{j=1}^{M_q} \sum_{i=1}^{R} (z_{ij} - c_i)^2 \right] / M_q$$

where c_i is the mean of the i-th dimension and M_q is the number of prototypes in class q.

(8) The data x_i for the classified object $\underline{x}$ are auto-scaled, transformed and normalized to give data t_i using the autoscaling, transformation and normaliz-ation parameters calculated for the class q in steps (1), (4) and (6).

(9) As a measure of similarity of the object $\underline{x}$ to the corresponding class q average squared distances of the object from all prototypes in this class are calculated

$$D_q = \left[\sum_{j=1}^{M_q} \sum_{i=1}^{R} (z_{ij} - t_i)^2 \right] / M_q$$

Alternatively, this similarity may be calculated on the basis of the following equation, which is more convenient for computational purposes

$$D_q = D_q^{(c)} + \sum_{i=1}^{R} (c_i - t_i)^2$$

(10) For comparability with other classes of the system, D_q are divided by the corresponding mean distances

$$D_q / D_q^{(c)} = \left[D_q^{(c)} + \sum_{i=1}^{R} (c_i - t_i)^2 \right] / D_q^{(c)}$$

or

$$D_q / D_q^{(c)} = 1 + \sum_{i=1}^{R} (c_i - t_i)^2 / D_q^{(c)}$$

Evidently, $D_q^{(c)}$ is a constant for a given class.
The first term, being equal to unity, can therefore
be neglected in the following comparative calculation. The measure of similarity S_q can then be
expressed in the form

$$S_q = \sum_{i=1}^{R} (c_i - t_i)^2 / D_q^{(c)}$$

Thus the measure of similarity in the SPHERE method
is the ratio of the squared distance of the object
from the centre of a class and the average squared
distance of this centre from all prototypes in the
class.

(11) For a dichotomy case, the object $\underline{x}$ is classified
into class q_1 if $S_q(1) < S_q(2)$ and into class q_2
if $S_q(1) > S_q(2)$. Generally, the object in a system
with Q classes is classified into class q with the
lowest value of S_q in comparison with those for
Q - 1 remaining classes.

(12) The relative similarity S^+ of the classified object
to the classes of the system can be expressed as
the ratio

$$s^+ = s_q'/s_q$$

where s_q' is the measure of similarity for a less
similar class and s_q is the measure for a more
similar class. For this ratio the condition $s^+ \geq 1$
is valid. For $s^+ = 1$ no decision is possible.
In general, the higher is s^+ the easier is the
decision. The acceptable value in a real problem
is estimated arbitrarily.

The SPHERE classification method determines the factors
in the class model on the basis of the solution of an
eigenvalue problem by means of the Karhunen-Loève transformation, i.e., by an approach which closely resembles
the principal component analysis in its "classical" version (see Sub-section 2.3.1). The number of factors in
the model that represents sufficiently the class is
estimated by a heuristic approach related to classification results. On the other hand, in the SIMCA method the
factors (components, latent variables) are calculated by
the NIPALS procedure (see the SIMCA method) and the number
of them is estimated by a cross-validation procedure
(Wold, 1978). The first method thus represents a "step-
down" method, whereas the SIMCA approach is a "step-up"
method of class modelling by means of factors.

In the SPHERE method the number of factors is estimated
by a criterion based on the minimum value of $\lambda^{1/2}$, which
is used in the normalization step (6) of the above SPHERE
algorithm (Fusek and Štrouf, 1979). It is determined from
the classification of the given system using a set of
different $\lambda^{1/2}$ values by interpreting the behaviour of
performance measures, viz., recognition rate and measure
of reliability (Fusek and Štrouf, 1979). These measures

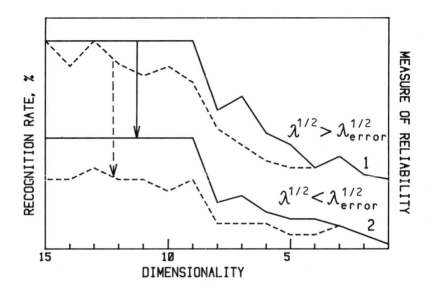

Fig.19. A typical graph for $\lambda^{1/2}_{error}$ estimation in the SPHERE method.
————— recognition rate,
------ measure of reliability.

do not vary significantly in the range of higher values
of $\lambda^{1/2}$ up to a certain value when the recognition rate,
as well as the measure of reliability, drops down (Fig.19).
This value is then defined as the $\lambda^{1/2}_{error}$, i.e., a standard deviation which should be ascribed to errors (noise).
The variables possessing the $\lambda^{1/2}$ lower than the $\lambda^{1/2}_{error}$
are deleted from the modelling, because the presence of
such factors completely deteriorates the modelling by
over-weighting the errors. The method has relatively high
requirements on computation, but it gives results accounting for the effect of errors on classification. For a set
of variables used in categorization of complex hydrides
(Section 1.2) the value of 0.12 has been found for the

$\lambda_{error}^{1/2}$. Also in the physical modelling of catalytic activity by variables of fundamental physico-chemical parameters this value has been successfully used (Section 1.2).

In 1979, the SPHERE method was first used in a classification of ABH_nD_{4-n} complex hydrides as stable or unstable compounds; the hydrides were characterized by physico-chemical parameters for atoms A (1st ionization energy) and B (melting point, density) and by structural descriptors for ligand D (molecular weight of the ligand, number of chain atoms, number of substituents on the first ligand atom, electronegativity of the 1st atom and π-donor or acceptor character of substituent) (Štrouf and Fusek, 1979; Fusek and Štrouf, 1979).

More recently, the behaviour of metals in some heterogeneous catalytic processes or in processes underlying certain catalytic reactions has been modelled. It has been found that the characterization of the metals by their fundamental physico-chemical parameters is suitable in such a SPHERE modelling as in, for example, the modelling of catalytic activity of transition metals in hydrogenolysis of ethane (Štrouf et al., 1981a). Furthermore, chemisorption and dissociation of carbon monoxide (Štrouf et al., 1981b) and chemisorption of hydrogen (Kuchynka et al., 1981) on metals have been modelled by the SPHERE method as initial basic steps in catalytic Fischer-Tropsch synthesis (see Section 1.2).

2.1.4 DEVIATION-PATTERN RECOGNITION

The deviation-pattern recognition uses nonlinear regression deviations (visualized as deviation plots) calculated for a known theoretical relation of system behaviour. This method is, therefore, applicable only

in cases when the corresponding equations are available.
It has been tested in two main situations:

(a)   The first situation is the testing of the validity of
       the regression model by simply looking at the devi-
       ation plot: if the plot is random the hypothesis is
       accepted, if the plot is nonrandom the hypothesis is
       rejected.

(b)   More sophisticated is an automatic comparison of the
       obtained deviation plot with the constructed
       (simulated) deviation plots (patterns). Such an
       automatic classification by deviation-pattern
       recognition seems to be promising but it is still
       in its infancy (Rusling, 1984).

The main recent applications are from electrochemistry.
The method has been used e.g. for resolution of severaly
overlapping signals and for elucidation mechanisms of
electrochemical reactions. In the latter case, Rusling
(1983) has used deviation-pattern recognition as binary
decisions for branch points in hierarchical tree, this
computerized method being suitable for mechanistic classification of one-electron potentiostatic current-
potential curves for a variety of working electrodes and
experimental conditions (Rusling, 1983a).

Meites (1982, 1982a) has recently employed the deviation
-pattern recognition approach for detection of the presence of an additional reaction component. For instance,
the method has been used in the kinetics of a pseudo-
first order reaction to differentiate between one and
two reactant systems by comparing the deviation plot for
a single reactant with that when a second reactant is also
present (Meites, 1982a). Similarly, the deviation-pattern
recognition can detect the presence of the second acid
and evaluate its concentration from deviations from the
measured and calculated curves for the potentiometric acid
-base titrimetry of a monobasic weak acid (Meites, 1982).

2.1.5   EVALUATION OF CLASSIFIERS

The quality of pattern recognition results must be carefully and critically checked after completing an analysis.
A variety of different approaches using more or less
sophisticated criteria of "goodness" of the classification
are used.

A conceptually simple approach is to divide the objects
of known class membership in two groups. The first group
is used for training of the classifier and is usually
called the training (learning) set (Section 2.1). The
second group serves for direct testing of the classifier
performance and is commonly called the test or prediction
set. Using the learning set one can, by means of an appropriate criterion, evaluate the quality of recognition
(that means to evaluate classification of the objects
with known membership to the classes used in the training
- prototypes). The recognition performance only indicates
how the classifier was adjusted and cannot be generalized
for the evaluation of the classifier from the point of
view of its prediction performance. The recognition rate
is generally over-optimistic and its use in connection
with prediction reliability cannot be recommended. A more
realistic evaluation of prediction performance of the
classifier can be achieved by using the test (prediction)
set if it is sufficiently representative.

A commonly used measure for the performance of classifiers is the percentage of correctly classified test
patterns (Varmuza, 1980, p.118). Unfortunately, this
criterion suffers from serious limitations. In particular,
very different populations in the test set for classes
q_1 and q_2 can yield nonobjective characterization of the
classifier.

Improvement consists of the determination of predictive
abilities P_1 and P_2 for each class separately

$$P_1 = p(q_1,y)/p(q_1) - p(y|q_1)$$

$$P_2 = p(q_2,n)/p(q_2) - p(n|q_2)$$

where $p(q_1,y)$ and $p(q_2,n)$ are the probabilities that the binary classifier answer "yes" if the object belongs into class q_1 and "no" if it belongs into class q_2, respectively, and $p(y|q_1)$ and $p(n|q_2)$ are the respective conditional probabilities.

For an equally populated prediction set for class q_1 and for class q_2 a simple criterion can be constructed as the average predictive ability $\bar{P}$ (Varmuza, 1980, p. 122).

$$\bar{P} = (P_1 - P_2)/2$$

Leave-n-out method

A relatively unbiased and therefore commonly used method for the evaluation of prediction ability is the approach called the "leave-n-out" method. It is the only one which operates even in situations when the test set is not available. In the simplest case, only one prototype ($n = 1$) is deleted from the training process to be used for testing a trained classifier. All M prototypes are subsequently treated in the same way by this leave-n-out procedure. The method is suitable for the detection of a highly dissimilar prototype (outlier). Nevertheless, a combinatory character of this procedure causes high requirements on computation. A compromise is an application of the leave-n-out procedure when $n > 1$. In each step a combination of n prototypes is deleted in such a way that each prototype is treated only once.

Direct comparison of classifiers

A direct comparison of pattern recognition classifiers can be used for evaluation purposes. Thus, Sjöström and Kowalski (1979) have described a comparison of five pattern recognition methods based on the classification results obtained for six sets of various types of data. Recently, results of the SIMCA method (Štrouf and Wold, 1977) and of the SPHERE method (Fusek and Štrouf, 1979) as obtained for the classification of complex hydride stability have been directly compared.

The direct comparison is independent of measures of performance; nevertheless, it is necessary to take into account that it is valid only for the studied system. Therefore, the results of direct comparison cannot be used as a general measure of "goodness" of the classifiers. On the other hand the direct comparison is a very important test from the point of view of the assessment of classification reliability. When the results obtained by different classifiers for a given system are sufficiently similar, then one can take the pattern analysis to be reliable to a reasonable extent. However, when the results differ significantly, the analysis is dubious. Therefore, for pattern recognition analysis of the given system the use of several different classifiers should be recommended.

Information theory

Information theory provides useful tools for evaluation of classifiers (Rotter and Varmuza, 1975; Varmuza and Rotter, 1976, 1978; Varmuza, 1980, pp.129-136). The terms of information theory are discussed in books (see, e.g., Eckschlager and Štěpánek, 1979, 1985) and they are used here without definitions.

The difference of uncertainties before and after classification H(A) and H(A|B) shows the decrease of uncertainty R and it is a form of an information measure called transinformation (Fig.20):

$$R = H(A) - H(A|B)$$

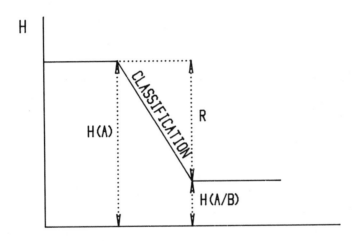

Fig.20. Graphical representation of transinformation.

The transinformation is expressed in probabilities; thus, a priori uncertainty regarding class membership in A is

$$H(A) = -p(1)ld\left[1/p(1)\right] - p(2)ld\left[1/p(2)\right]$$

in which p(1) is the a priori probability of membership in A and p(2) = 1 - p(1) is that of membership in B (complement of A).

The transinformation is dependent on the population of the prediction set. Therefore, the "figure of merit" M

$$M = R/H(A)$$

has been introduced which enables a valid comparison of different classifiers applied even to differently composed test sets (e.g. Lam T.F. et al., 1976).

For classifiers with continuous response, overlapping classes can be a serious question, which must be considered in the evaluation. In the case of such classes, the Bhattacharyya coefficient B or the transinformation for continuous distributions should be used (Varmuza, 1980,pp.136-138).

Recently, the information theoretic approach has been used to evaluate confidence in the results of learning machines trained on mass spectra (Richards and Griffiths, 1979). Most recently, Wienke and Danzer (1985) have proposed a new measure for the evaluation of classification based on a combination of information theory formulas and geometrical principles. The method is useful predominantly in the evaluation of different classification algorithms applied to the same data base or, on the other hand, of the same algorithm applied to different data bases.

2.2 UNSUPERVISED CLASSIFICATION METHODS

Unsupervised recognition methods (nonsupervised learning, recognition without supervision, learning without teacher, numerical taxonomy, mode seeking etc.) have a single goal - to discover significant clusters of patterns in D-dimensional space. When $1 \leqq D \leqq 3$, then the human inspection is a very powerful tool for recognition of clusters.

For $D > 3$ two main approaches can be used:

(a) reduction of the dimensionality to "human" dimensionality $1 \leqq D \leqq 3$ to display the regularities in the system to be visualized (Section 2.4) or

(b) mathematical cluster analysis in multidimensional space, which can be, similarly to supervised procedures, divided into nonparametric (distribution free) methods and parametric methods (Andrews, 1972, pp.141-168).

2.2.1 CLUSTER ANALYSIS

Cluster analysis is a less frequently used method in analysis of chemical data than supervised pattern recognition methods, in spite of its fundamental importance for an unbiased insight into the data structure. In our opinion, the cluster analysis method of pattern recognition should be used in an iterative manner before implementing a supervised method.

The cluster analysis procedure can be, in principle, based on agglomerative or divisive methods, the former method being more frequently used. This method sequentially merges the objects of the system into clusters according to some rules. The rules for clustering are generally of a heuristic nature and, therefore, the meaningful result of the cluster analysis is a single criterion for the evaluation of a selected rule.

Nevertheless, the concept of similarity is a general basis of all clustering methods, irrespective of the selection of the similarity measure and the used algorithm. The similarity is usually measured in some distance units.

The most used (and for people most familiar) is the Euclidean distance d_E

$$d_E = \left[\sum_{i=1}^{N} (a_i - b_i)^2 \right]^{1/2}$$

in which a_i and b_i are the values of the i-th variable for the points a and b, respectively, and N is the number of variables.

Other distance measures can be derived from the general Minkowski distance d_M

$$d_M = \left[\sum_{i=1}^{N} (a_i - b_i)^k \right]^{1/k}$$

For example, k = 1 for the Manhattan (city block) distance (see Varmuza, 1980, pp.25,26).

After the selection of an appropriate distance measure, a distance matrix is calculated. In other words, the mutual distances between all possible pairs of objects are determined and stored in an ordered (matrix) form. Then the distances are ranked according to their increasing values. In each step, the minimum distance between two points (objects) is sought for and the corresponding pair is merged to one point (usually to a centre point)(Fig.21). This procedure is used in agglomerative hierarchical clustering.

Hierarchical clustering

Agglomerative hierarchical (Q-mode) clustering is a process in which the clusters are stepwise fused up to a number determined either on the basis of the chemist's a priori knowledge or by some rule. Such a rule for

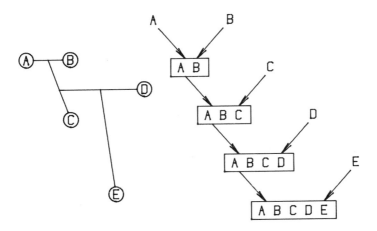

Fig.21. Scheme of hierarchical agglomerative clustering.

stopping the hierarchical agglomeration can be, principally, of two types:

(a) In the following step of the procedure the shortest
     distance exceeds a value fixed in advance or
(b) in a relative way, the increase of the minimum
     distance surpasses a given limit.

   Instead of the direct use of a distance some similarity
measures based on the distance can be employed. For
example, the similarity $S(\underline{a},\underline{b})$ between patterns $\underline{a}$ and $\underline{b}$
can be calculated according to

$$S(\underline{a},\underline{b}) = 1 - d(\underline{a},\underline{b})/d_{max}$$

where $d(\underline{a},\underline{b})$ is the Euclidean distance between the points
(objects) $\underline{a}$ and $\underline{b}$, and d_{max} is the largest distance found

in the given data set (Kowalski and Bender, 1972a). The
matrix based on the similarities is called the equality
matrix (Kateman and Pijpers, 1981,p.211).

The results of hierarchical clustering are advantageously represented by the so called dendrogram, or classification trees, as illustrated in Fig.22.

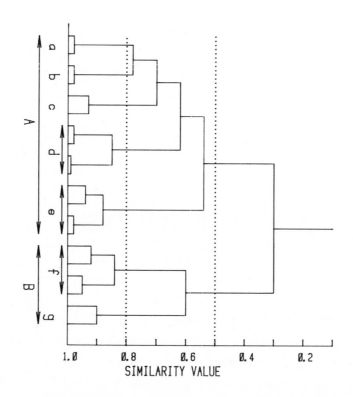

Fig.22.   Hierarchical clustering dendrogram.
           When similarity value of 0.8 is selected as a
           limit for "significance" of clusters, seven
           clusters (a-g) can be found; when this value
           equals 0.5, then only two clusters (A,B) can
           be detected.

Fractal or 3-distances clustering method is a new hierarchical procedure proposed by Zupan (1980, 1982, 1982a) for updating large data bases organized as binary trees. The method is based on the calculation of three distances at any given vertex l on the level n, $\underline{a}(l,n)$ instead of two distances only, as usual in classical methods. The respective distances $d_1(\underline{x},\underline{a}_l)$, $d_2(\underline{x},\underline{a}_r)$ and $d_3(\underline{a}_l,\underline{a}_r)$ are calculated between the input $\underline{x}$ and the left and the right descendant of the vertex $\underline{a}(l,n)$, $\underline{a}_1$ and $\underline{a}_2$, respectively. The main advantage of this new Zupan method compared to the standard clustering methods (formation of hierarchical trees) is that the required memory or computational time is reduced from M^2 to approximately MlogM (M is the number of clustered objects). This advantage opens the possibility to use hierarchical clustering even for relatively large data bases as has been shown for 500 infrared spectra library of polymer compounds (Zupan, 1982).

The agglomerative hierarchical methods reduce the matrix one unit at a time and, therefore, clustering programs have the reputation of needing a lot of computer memory and time. Kaufman et al. (1983) have examined the possibility of using microcomputers for clustering by employing Macnaughton-Smith's algorithm (Macnaughton and Smith, 1964), which belongs to divisive hierarchical types of algorithms. This means that one starts out with all the objects to be clustered and divides them first in two groups; each of these groups is then further divided and so on, as is shown in Fig.23.

Non-hierarchical clustering

Massart and co-workers have developed a new cluster analysis technique based on the p-median non-hierarchical clustering method (Massart et al., 1980, 1981, 1981a,

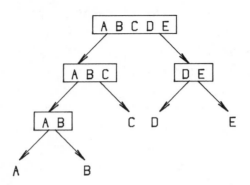

Fig.23.   Scheme of hierarchical divisive clustering.

1982b, 1983, 1983a; Massart-Leën and Massart, 1981; Esbensen, 1984). The program package has been named MASLOC. The method, originating from location theory, selects a given number p of objects in such a way that the sum of distances of other objects to the closest of the p objects is minimized. Moreover, each object is simultaneously assigned to one of p selected objects in a clustering process.

Two groups of binary variables are necessary to describe mathematically the location model (Massart et al., 1980, 1983):

(a)   a group of variables denoting the selection of p representative objects, y_j, which is equal to one if object $\underline{x}_j$ (j = 1,2,...,M; M is the number of objects) is selected as a representative object and to zero if not, and

(b)   a group of variables describing the assignment of objects to clusters, x_{jk}, which is equal to one if an object $\underline{x}_k$ is the closest to the representative object

$\underline{x}_j$ and is therefore assigned to the cluster that is represented by $\underline{x}_j$, and to zero if not.

The model can be then described as follows: minimize the function $f(p)$

$$f(p) = \sum_{j=1}^{M} \sum_{k=1}^{M} d_{jk} x_{jk}$$

in which d_{jk} represents the distance between the objects $\underline{x}_j$ and $\underline{x}_k$ in a commonly used distance measure (see Subsection 2.2.1) under the constraints

$$\sum_{j=1}^{M} x_{jk} = 1$$

$$x_{jk} \leq y_j$$

$$\sum_{j=1}^{M} y_j = p$$

$$y_j, x_{jk} \in \{0, 1\}$$

The optimum solution can be found

(a) by an exact branch and bound algorithm (Koontz et al., 1975; Diehr, 1985) for data sets of a moderate size or

(b) by heuristic algorithms for large data sets.

The method eliminates the rigidity of clusters in frequently used agglomerative hierarchical methods. This means that this novel non-hierarchical clustering method does not enforce the elements of a formed cluster to

remain together in all subsequent steps of the clustering
and thus it is not influenced by possible incorrectness
in the early stages of the clustering process.

This Massart clustering method, moreover, permits one
to determine
(a)   outliers (clusters with only one member at low p),
(b)   robust, "significant", clusters (those with elements
       that do not intermingle with elements from other
       clusters at higher p), and
(c)   tight clusters (those with elements that keep
       together until a high p value is reached (Massart
       et al., 1982b).

Finally, this clustering process can be advantageously
represented by a hierarchical plot (MASLOC dendrogram) in
a "hierarchical—non-hierarchical clustering strategy"
used by Massart et al. (1982b).

Potential clustering method CLUPOT

Coomans and Massart (1981a) have described a new clustering procedure named CLUPOT suitable for two types of
application:
(a)   the detection of subsets (clusters) of interrelated
       objects and
(b)   the selection of small number of objects highly
       representative for the classes (centrotypes).

The potential surface is constructed as in the ALLOC
method (Sub-section 2.1.2) by averaging individual
Gaussian potential functions centered in points (objects).
But contrary to the procedure in the supervised method
ALLOC, in this unsupervised case the calculation is
carried out over the whole data set. The peaks on the
potential surface indicate the presence of clusters. The
object in a cluster that is situated in the place with

maximum average potential is called the cluster centro-
type. In the CLUPOT method, the centrotypes found are then
used in clustering the objects.

As in all cluster analysis methods, the determination
of a correct number of significant clusters is a serious
problem also in CLUPOT, because of influence of a smoothing parameter α used. Different values of this parameter yield various numbers of peaks on potential surface
(i.e. numbers of clusters, no.c.). The concept of reliability curves has been applied for the determination of
the number of significant clusters. If a stepwise increase
of the value of the smoothing parameter does not change
the number of clusters, then the clusters are significant
(Fig.24).

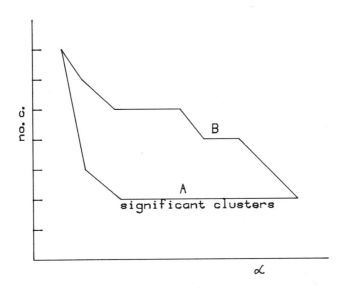

Fig.24. Curve of reliability in the CLUPOT method.
Curves A and B correspond to reliable and
unreliable clustering, respectively.

2.2.2 MINIMAL SPANNING TREE

Minimal Spanning Tree (MST) algorithms (Zahn, 1971) start
with points (objects) connected in such a manner that the
total length is minimum (MST) of all possible combinations
(circuits are excluded). Then the algorithm searches for
the largest distance and interrupts it. Thus two clusters
are formed in which the algorithm finds the next largest
value and breaks it and so on until a given condition is
fulfilled (Varmuza, 1980,pp.95,96).

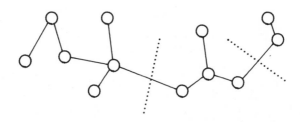

Fig.25. Minimal Spanning Tree.

   The cluster analysis approach has been recently used in:
- classification of chemical structures of benzenoic compounds represented by automatically derived structural
  features (Adamson and Bawden, 1981),
- trace element data analysis (Pillay and Peisach, 1981;
  Arunachalam et al., 1983),
- simplification of the computerized identification of
  compounds on the basis of their mass spectra (Domokos
  et al., 1980),
- characterization and selection of stationary phases for
  gas chromatography (Huber and Reich, 1984),

- structure-property (Willett, 1982a) and structure-
  activity (Chen et al., 1979; Takahashi et al., 1980;
  Miyashita et al., 1981a) relationships studies,
- modelling of optimum component combinations in samples
  from clinical chemical laboratory (Goldschmidt et al.,
  1983),
- classification of geochemical organic samples (Vuchev,
  1983) and
- examination of chromatographic profiles of urine
  proteins (Marshall et al., 1984).

2.3 FEATURE SELECTION

Feature selection is a very important part of each pattern
recognition analysis (see e.g. a special issue of IEEE
Trans. Comput., 1971). Unfortunately, no general theoretical basis is available for this process. The selection of
features is thus highly dependent on the a priori knowledge of an interpreter. The mathematics can only substi-
tute the interpreter in processing the data in multidimensional space with the aim of providing an insight into
their structure. The validity of feature selection can be,
therefore, checked only by classification results. Many
feature selection methods are heuristic to a high extent,
as they are controlled mainly by the classification performance.

The single unbiased way to select an optimum subset of
features from the given original feature set is to compare
the classification results for all possible combinations
of features. It is evident that even for high-speed computers such an approach requires an unrealistic computational burden even for relatively small sets of data.
Therefore, some more economical, albeit suboptimum, procedures should be used.

In this book, the term "feature selection" means to search for linearly independent variables that are relevant in respect of the classification problem under consideration. This definition of feature selection, limited to some extent, has a single goal: to find the so called "intrinsic dimensionality" of the system (or in an approximation of its model)(Štrouf and Fusek, 1979).

It should be mentioned that the extraction of features is often incorrectly connected with feature selection in spite of the fundamentally different characteristics of both processes. The former process tries to represent a set of given objects in terms of patterns. That means, in this stage of pattern recognition analysis, the objects are represented by strings (vectors) of data and a table of these data can be formed. In some cases the data are available in a form suitable to be tabulated directly (e.g. the results of a chemical analysis, the values of physico-chemical parameters, the codes for presence or absence of a structural characteristic and so on). On the other hand, there are numerous cases when a continuous graph or image must be represented with a sufficient number of digital data: this process is called here feature extraction. For the extraction of features from some special types of graphs (for example, from spectra, from experimental curves, etc.) there exists specific experience, and the optimum approach may be found from the original literature studying the given type of graphs. In other words, the feature extraction is, in principle, a formation of the table as defined in Chapter 1; in this book it is not treated in detail.

Feature selection, in contrast to feature extraction, possesses an analytical character. In one step it searches for the interrelation between variables and in a second step for their classification relevance. In the former case a dimensionality reduction process takes place, whereas in the latter one the deletion of redundant (non-relevant) variables is carried out.

2.3.1 DIMENSIONALITY REDUCTION

The dimensionality reduction is made mostly by linear de-correlation. For this purpose, an appropriate transformation can be suitable. The most frequently used rotational transformation applied for searching for linearly independent variables is principal component (factor) analysis, which can be optimally implemented by Karhunen-Loève transformation (rotation, expansion). This transformation is based on the eigenvector problem (Andrews, 1972, p.27)

$$\underline{U} \cdot \underline{e}_r = \lambda_r \underline{e}_r$$

where $\underline{U}$ is the covariance matrix for prototypes, $\underline{e}_r$ is an eigenvector of the covariance matrix $\underline{U}$ and λ_r is the r-th corresponding eigenvalue which gives the minimum estimation error, E_N

$$E_N = \sum_{r=N+1}^{R} \lambda_r$$

in which λ_r, $r = N+1, \ldots, R$, are the eigenvalues associated with those eigenvectors not included in the expansion of equation

$$z_j = \sum_{r=1}^{N} a_{jr} e_r$$

The error E_N will be minimum if the first N feature eigenvectors are chosen in such a way that they correspond to
the largest eigenvalues. The problem of the minimization
of the error E_N is the basis of principal component
analysis.

Other orthogonal transformations, e.g. Fourier, Walsh
and Haar transformations, have been investigated as
methods for feature selection. The theoretical considerations have been tested by results obtained by different
classification methods for mass spectral data sets
(Domokos and Frank, 1981).

The principal component (factor) analysis has been
recently used in various fields of chemistry and some of
them are mentioned here:
- Non-random part of the variability of repetitive pyrolysis-gas chromatograms can be modelled by separate
  principal component models (Blomquist et al., 1979).
- Factor analysis facilitates the formulation of a satisfactory correlation equation for QSAR (Chen B.K. et al.,
  1979).
- Principal component analysis has been used to decompose
  pharmacological activity indices of chemical compounds
  into mutually independent components (Lukovits and
  Lopata, 1980; Lukovits, 1983).
- Factor analysis can be used for the characterization of
  the chemical state from line shapes of the Fourier-
  transformed X-ray-excited carbon Auger spectra as demonstrated on two systems (Gaarenstroom, 1979).
- Principal component analysis has been used to investigate correlation between objective chemical measurements
  and subjective sensory evaluation by a panel of judges
  in the quality evaluation of wines (Kwan and Kowalski,
  1980).
- Principal component analysis has been used for evaluation of the relation between quantitative variables

and chemical terms in acridine derivatives (Mager, 1980).

- Principal component analysis was employed in the multi-variate analysis of ionization constant data (Edward et al., 1981).

- Factor analysis of mass spectra from partially resolved chromatographic peaks was evaluated on simulated Gaussian data with and without imbedded errors (Woodruff et al., 1981a).

- Principal component analysis has been applied to ^{13}C-NMR chemical shift assignment of chalcones and their thiophene and furan analogues (Musumarra et al., 1981) and acrylonitriles derivatives (Musumarra et al., 1981a) as well as to ^{13}C and ^{15}N chemical shifts of triazenes (Dunn et al., 1982a).

- Factor analysis has been used for reduction and ranking of capillary gas-chromatographic data from an analysis of the cigarette smoke (Hsu et al., 1982).

- Principal component analysis can compare composition and structural parameters of heavy oils on the basis of analytical data (Matsubara, 1982).

- Principal component analysis has been reported for geo-chemical and geological applications (Esbensen and Wold, 1983).

- Principal component analysis and factor analysis have been used in organic geochemistry (Vuchev, 1983).

- Principal component analysis can evaluate complex mixtures of non-volatile compounds characterized by their mass spectrometric data (van der Greef et al., 1983).

- Principal component analysis and factor analysis (followed by discriminant analysis and graphical rotation) can interpret partial chemical composition of complex biological samples on the basis of their pyro-lysis-mass spectra (Windig et al., 1983).

- Factor analysis has been used to distinguish between several cigarette types on the basis of glass capillary

gas-chromatographic data for the organic phase of the
respective cigarette smoke (Parrish et al., 1983).
- Principal component technique in combination with Fisher
discriminant method (see Sub-section 2.1.1) can be used
for the feature selection for ^{13}C-NMR spectra
(Arunachalam and Gangadharan, 1984).
- Principal component method enables component spectra
from pigment mixtures to be estimated (Meister, 1984).
  Moreover, as a curve resolution technique, principal
component method has been used in the analysis of emission
(Aartsma et al., 1982) and fluorescence (Saltiel and
Eaker, 1984) spectra as well as in the analysis of
potentiometric data (Havel and Meloun, 1985).

2.3.2 RELEVANCE OF VARIABLES

The relevance of variables in pattern recognition consists
of two abilities of the variables:
(a) the ability of a variable to cluster the objects
    together and
(b) the ability to distinguish between classes.
  The variables with both high abilities are considered
to be relevant in a given classification analysis. The
common measures that relate both abilities in the variance scale are the Fisher's weight and the variance weight.

Weighting of variables

The weighting is a multiplication of original variables
by coefficients which express the importance of individual
variables in the problem under consideration (Varmuza,
1980,p.107; Kateman and Pijpers, 1981,p.213).

Fisher's weighting

The Fisher's weighting of a variable x_i (Fisher, 1936) belongs to frequently used discriminatory weighting methods (Pijpers et al., 1979; Varmuza, 1980,p.108), in which the weight for the i-th variable can be expressed as (Kateman and Pijpers, 1981,p.214)

$$w(Fisher)_{i,k,l} = (\overline{x'_{i,k}} - \overline{x'_{i,l}})^2 / M_k \, \sigma^2_{i,k} + M_l \, \sigma^2_{i,l})$$

where $\overline{x'_{i,k}}$ and $\overline{x'_{i,l}}$ are the means for the i-th autoscaled variable in classes q_k and q_l, respectively, M_k and M_l are the numbers of prototypes in these classes and $\sigma^2_{i,k}$ and $\sigma^2_{i,l}$ are the variances for the respective classes and for the variable x_i. This weighting method was modified by Coomans et al. (1978) to a computationally more convenient form.

The Fisher's weighting factor of a variable x_i for all linear class separations can be expressed as (Pijpers et al., 1979)

$$w(Fisher)_i = 2 \left[\sum_{k=1}^{Q-1} \sum_{l=k+1}^{Q} w(Fisher)_{i,k,l} \right] / \left[Q(Q-1) \right]$$

Variance weight

The variance weight of a variable x_i is determined as the ratio of interclass variance and intraclass variance of the autoscaled variable x'_i (Kowalski and Bender, 1972a; Pijpers et al., 1979).

$$w(var)_{i,k,l} = \left[\overline{(x_k')_i^2} + \overline{(x_l')_i^2} - 2\overline{(x_k')_i}\,\overline{(x_l')_i} \right] /$$
$$/ \left[(m2)_{i,k} + (m2)_{i,l} \right]$$

where $(m2)_{i,k}$ and $(m2)_{i,l}$ are the second central moments of the i-th variable in classes q_k and q_l

$$(m2)_{i,k} = (M_k - 1)\,\sigma_{i,k}^2 / M_k$$

$$(m2)_{i,l} = (M_l - 1)\,\sigma_{i,l}^2 / M_l$$

where M_k and M_l are the number of prototypes in classes q_k and q_l.

The variance weighting of variables x_i for all linear class separations is

$$w(var)_i = \left[\prod_{k=1}^{Q-1} \prod_{l=k+1}^{Q} w(var)_{i,k,l} \right]^{2/[Q(Q-1)]}$$

Generally expressed, the aim of all weighting methods is to rank variables according to their importance in the given classification problem. Three basic situations should be then considered:

(a) All weighted variables are used in the following analysis - the preprocessing is carried out (Section 3.1).

(b) Only several (but more than three) variables with the largest weights are retained in the analysis - the procedure belongs to feature selection methods (this Section).

(c) Finally, only three or fewer variables from the most
relevant variables are utilized - the visualization
is carried out (Section 2.4).

A two-criterion approach has been proposed by Wold (see
Wold and Sjöström, 1977; Sjöström and Wold, 1980) in
connection with his disjoint class modelling by the SIMCA
method (Sub-section 2.1.3). In first step, the modelling
power $\Psi_i^{(q)}$ is determined, which is connected with the
ability of the variable to model the class q as measured
in terms of standard deviations

$$\Psi_i^{(q)} = 1 - (RSD_i^{(q)}/SD_i^{(q)})$$

where $RSD_i^{(q)}$ is the residual standard deviation of variable x_i over all the data in the training set and $SD_i^{(q)}$
is the standard deviation of the training set data.
A value of $\Psi_i^{(q)}$ close to one indicates a high modelling
power. In second step, the discriminatory power $\phi_i^{(r,q)}$
is calculated in the following way: the objects belonging
to class r are fitted to the model for class q and vice
versa. The residual standard deviations $RSD_{i,r}^{(q)}$ and $RSD_{i,q}^{(r)}$
are compared to those when the objects are fitted to
their "own" classes, $RSD_{i,r}^{(r)}$ and $RSD_{i,q}^{(q)}$

$$\phi_i^{(r,q)} = \left[(RSD_{i,r}^{(q)2} + RSD_{i,q}^{(r)2})/(RSD_{i,r}^{(r)2} + RSD_{i,q}^{(q)2}) \right]^{1/2} - 1$$

A high value of $\phi_i^{(r,q)}$ indicates high discriminatory
power of the variable x_i. Only the variables with high
values for both modelling and discriminatory powers are
retained in the pattern recognition analysis (see e.g.
Štrouf and Wold, 1977).

Information content of a variable

The information amount $I(S,x_i)$ extracted from a given variable (dimension) can be expressed by the difference of total source entropy $H(S)$ and entropy of the source given a feature value $H(S|x_i)$ (Cleij and Dijkstra, 1979; see Eckschlager and Štěpánek, 1979,p.157, 1985,p.20)

$$I(S,x_i) = H(S) - H(S|x_i)$$

where S is a source and x_i is a value of the i-th dimension. This information obtained from the i-th variable is a measure of relevance of the corresponding coordinate i. Only a certain number of dimensions i with the largest information amounts $I(S,x_i)$ is used in further analysis.

Feature selection with ALLOC

Coomans et al. (1981b) have introduced the feature selection procedure of supervised pattern recognition method ALLOC (Hermans and Habbema, 1976) (see Sub-section 2.1.2) for chemical applications. The procedure starts with the selection of the single variable which discriminates best between the training sets. Then all the variables are ranked in a stepwise way: the new selected variable in each following step must be discriminatory optimum in combination with the variables selected in the preceding steps. The discriminatory power of an individual variable for each combination in a given step is measured as the probability of error

$$P(error) = \sum_{k=1}^{Q} P(q)M_q(error)/M_q$$

in which $q = 1,2,...,k,...,Q$ represents Q learning sets
and $P(q)$ is the a priori probability of q usually
$P(q_1) = ... = P(q_k) = ... = P(q_Q) = 1/Q$; $M_q(error)$ is the
number of prototypes of class q misclassified in the
leave-one-out evaluation procedure (Sub-section 2.1.5)
and M_q is the number of all prototypes in class q. The
procedure is stopped either when all variables are ranked
or when a limit in the increase of the discrimination is
reached. The performance of this classification-dependent
feature selection has been compared with those of the
classification-independent ones, the SELECT procedure in
the ARTHUR package (Duewer et al., 1975; Kowalski and
Bender, 1976; Harper et al., 1977) and the procedure based
on F-test in the SPSS package (Nie et al., 1975).

Intrinsic dimensionality

Štrouf and Fusek (1979) have recently described an approach for determination of intrinsic dimensionality of
chemical models by the SPHERE pattern recognition method
(see Sub-section 2.1.3). First, the variables are ordered
according to their importance in respect to the given
classification of the objects under consideration. The
evaluation of the classification ability of the individual
variables may be performed by any appropriate method without considering, in this stage, whether the variables are
mutually linearly dependent or not. For example, one such
approach to the classification relevance evaluation can
be combined criterion using discriminatory and modelling
powers of the variables (see preceding paragraphs).

   Then, the determination of linearly independent variables is performed to estimate the minimum number of
mutually independent relevant variables which sufficiently
exactly define all objects in the system. This number is

a fundamental characteristic of each system (including
the chemical one) and can be, geometrically, represented
as an orthogonal basis with dimensionality equalling the
number of linearly independent relevant features. Thus
from the point of view of classification (and consequently from the point of view of pattern recognition), the
intrinsic dimensionality of the system, D^S, can be defined
as the minimum number of linearly independent discriminatory relevant variables (Štrouf and Fusek, 1979).

The determination of linearly independent variables is
based on the Karhunen-Loève transformation. New variables
are a linear combination of the original variables with
coefficients called eigenvectors. However, the physical
meaning of transformed variables in relation to the original ones is not explicitly known and they represent
abstract factors. The problem of a "come-back" into
measurement space is not trivial and it is solved, for
example, by target factor analysis (Howery, 1977;
Malinowski and Howery, 1980).

Štrouf and Fusek (1979) have described, in connection
with the intrinsic dimensionality estimation, a new
simple stepwise procedure based on the idea that the
deletion of the feature that corresponds to the largest
component of the eigenvector related to the lowest eigenvalue, λ_{min}, minimizes the loss of information (Fig.26).
Nevertheless, in the case of identical or very similar
values no unique decision is possible and alternative sets
of original variables should be considered for the chemical model.

It should be emphasized that the Karhunen-Loève transformation does not account for the discriminatory effect
of variables (Kittler, 1977) when they are treated in the
whole without considering the clustering effects. However,
when the minimum eigenvalue λ_{min} approaches zero, this
"insensivity" in respect to discrimination does not repre-

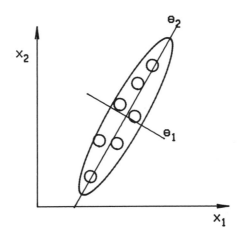

Fig.26. Deletion of variable in the SPHERE method.

sent any serious problem because the information (including the discriminatory information) of the deleted variable is preserved in the set of remaining features (Fig.
27). On the other hand, in the case when $\lambda_{min} \neq 0$ some
information is lost (Fig.28). Hence, for pattern
recognition purposes any additional unique deletion of a
variable is not possible using only the eigenvalue criterion. Nevertheless, a remarkable increase in the percentage loss of information

$$L(\%) = 100 \, \lambda_{min}^{(i)} / \sum_{r=1}^{R} \lambda_r^{(i)}$$

(where R is the dimension after i-th deletion step) may
be considered as a criterion for the determination of the
model intrinsic dimensionality D^m. The loss of discrimi-

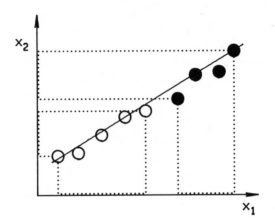

Fig.27. Discrimination of classes when $\lambda_{min} \doteq 0$.

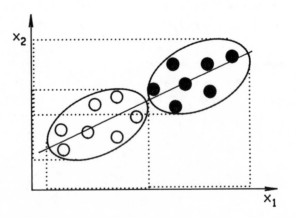

Fig.28. The loss of discriminatory information in the case when $\lambda_{min} \neq 0$.

natory information can be checked by classification performance using different evaluation measures (Sub-section 2.1.5) and various classification methods (Section 2.1).

The determination of intrinsic dimensionality by the Karhunen-Loève transformation (implemented computationally by the Jacobi procedure) is one of the main parts of the SPHERE pattern recognition polyalgorithm which has been tested by analysis of stability behaviour of complex hydrides and by modelling catalytic activity of metals (Section 1.2).

Recently, Danzer et al. (1984) have used another two-step feature selection in connection with Multivariate and Discriminant Analysis (MVDA). In the first step, the variables are ranked according to their influence on the classification. In the MVDA method (Läuter and Hampicke, 1973) only variables with small intraclass variance and with high discriminant power are retained. In the analysis of variance the variables are added step by step using an appropriate F test based on Hotelling's T^2 as a test statistic (see Dove et al., 1979). In the second step, the relevant variables are decorrelated by the solution of the eigenvalue problem (see preceding paragraphs). After deletion of highly linearly dependent variables, a set of relevant linearly independent variables is obtained.

Moreover, the two most relevant features z_1 and z_2 can be used for display (Fig.29), in which the radius r is calculated according to Ahrens and Läuter (1974)

$$r_q = \left\{ \left[T(M-Q)(M_q+1)F_{\alpha, T, M-Q-T+1} \right] / \left[M_q(M-Q-T+1) \right] \right\}^{1/2}$$

where T is the number of relevant features z_t (in Fig.29 T = 2) and F_α is a quantile of F-distribution for level of significance $\alpha = 0.05$. This approach was applied to

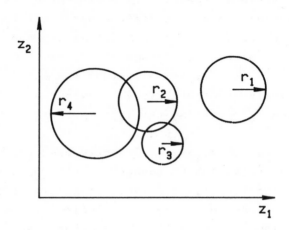

Fig.29. Graph of classes q_k (k = 1-4) in the space
defined by MVDA.

the classification of historical glass bead finds, using
a set of spectrographically determined concentrations of
colouring elements (Danzer et al., 1984). Furthermore, the
MVDA method has been advantageously used for the investigation of structural dependence of hyperglycemic and hypo-
glycemic activity of o-toluenesulfonylurea and thiourea
derivatives (Dove et al., 1979) and of homogeneity of
solids (Danzer and Singer, 1985). In the latter case, the
training set for class q consists of the values of the
repeated analyses in a given point of the solid surface.

2.4 VISUALIZATION

Various multivariate statistical techniques, including
pattern recognition techniques, can be used for the vis-

ualization of multivariate data which are not directly
accessible to human perception. The common goal of display
methods is to represent the data structure of points in
N-space by the same number of points in R-space where
$N > R$ and $1 \leqq R \leqq 3$, usually 2 (Kowalski and Bender, 1973).
Using computer graphics $R = 3$ can be also advantageously
displayed.

2.4.1 VARIABLE-BY-VARIABLE PLOTS

In the simplest case of variable-by-variable (Var-Var)
plotting there are $N(N-1)/2$ possible plottings and their
implementation is a cumbersome task; nevertheless, this
trivial, errorless visualization method can reduce significantly redundance in the data and it should be therefore carried out before the following parts of an analysis
(Fig.30a,b).

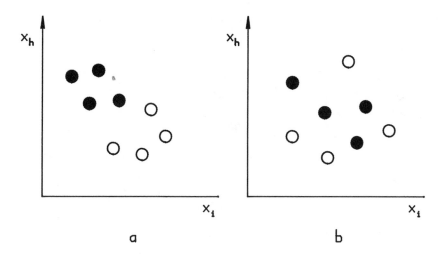

Fig.30. Var-Var plotting of two variables.
         a/   classes can be separated by two variables,
         b/   separation does not occur and more sophis-
               ticated projection should be examined.

A more sophisticated method is e.g. Var-Var plotting of linear combinations of the original variables, the rotation being controlled by an interactive computer graphics terminal.

2.4.2 LINEAR TRANSFORMATION

The linear methods are projections on new two coordinates which are linear combinations of N original measurements.

Eigenvector projection

The most commonly used linear method is eigenvector projection based on the solution of the eigenvalue problem (see Sub-section 2.3.1), which gives a set of N eigenvalues λ and a set of N eigenvectors $\underline{e}$; two eigenvectors $\underline{e}_1$ and $\underline{e}_2$ corresponding to two largest eigenvalues λ_1 and λ_2 are used as the coordinates of the projection plane (Fig.31).

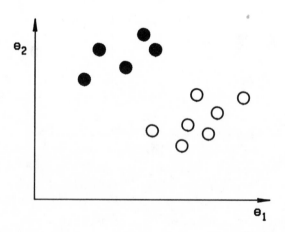

Fig.31. Eigenvector projection.

Eigenvector projection is an optimum linear projection
with minimum mean-squares error of variance. The advantage of this visualization method is that the relative
loss of information in terms of variance can be easily
calculated (Sub-section 2.3.1).

Linear projections from N-space to 2-space are called
by Wold et al. (1983b) "two-dimensional windows into
N-space". The authors have recommended for more complex
data sets, not only a plot of the first and the second
components, but also plots of the third and the fourth
largest components from the ordered set of the components.
In the SIMCA method (Sub-section 2.1.3) the components are
calculated by the least-squares method and the coefficients are called loadings. Generally, the linear projections are the most investigated, tested and applied display methods.

Fisher discriminant projection

Fisher linear discriminant functions have been used for
classification of mass spectral data in a two-level network of discriminants (Rasmussen et al., 1979a). The discriminants provide a useful basis for two-dimensional
projection.

2.4.3 NONLINEAR MAPPING

Nonlinear mappings are based on various criteria and are
often closely related to classification. Hence, there are
many different nonlinear mapping methods. The most commonly used is nonlinear mapping by error minimization
(Sammon, 1969; Kowalski and Bender, 1972a), which was
modified by Kowalski and Bender (1973). This modification

uses the Polak-Ribiere method (Polak, 1971, p.53) to minimize an error function E

$$E = \sum_{j<k} (d_{jk}^+ - d_{jk})^2 / (d_{jk}^+)^\varrho$$

in which d_{jk}^+ is the interpoint distance in N-dimensional
measurement space calculated according to

$$d_{jk}^+ = \left[\sum_{i=1}^{N} (x_{ij} - x_{ik})^2 \right]^{1/2}$$

and d_{jk} is the two-space interpoint distance in the
eigenvector plot (Sub-section 2.4.2). The value of ϱ is
selected such that it correctly weights small and large
distances; the value 2 weights the distances equally
(Kowalski and Bender, 1983).

The optimization procedure of non-linear mapping requires to compute $M(M-1)/2$ distances (M is the number of
objects) and thus it is very long even with a large computer. Therefore, Forina and Armanino (1982) have proposed a simplified procedure which uses, at its beginning,
the clustering idea of Crawford and Morrison (1968): the
autoscaled data x_{ij}' are normalized to generate points x_{ij}''
on the surface of a unity-radius hypersphere (r = 1)

$$x_{ij}''(q) = x_{ij}'(q) / \sum_{i=1}^{N} \left[x_{ij}'(q) \right]^2$$

and

$$\sum_{i=1}^{N} \left[x_{ij}''(q) \right]^2 = 1$$

Then N dimensions (coordinates) of the centre of gravity $\underline{c}^{(q)}$ of each cluster are computed

$$c_i^{(q)} = \sum_{j=1}^{M_q} x_{ij}^{(q)} / M_q$$

in which $c_i^{(q)}$ is the mean of the i-th variable in class q and M_q is the number of prototypes for this class. The distance of the centre of gravity $\underline{c}^{(q)}$ to the centre of the hypersphere is the Crawford-Morrison degree of clustering.

Instead of centre of gravity, Forina and Armanino (1982) used a cluster centre $\underline{p}^{(q)}$ as the projection of the centre of gravity onto the hypersphere surface (Fig.32).

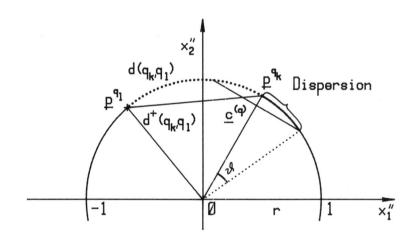

Fig.32.   Two-dimensional graph of hypersphere-type cluster analysis (for description see discussion in text).

Then an arbitrary position of the centre $\underline{p}^{(q_k)}$ is selected and the remaining centres $\underline{p}^{(q_l)}$ ($l \neq k$) are moved by changing their coordinates to minimize the error function

$$E = \sum_{\substack{k,l=1 \\ k \neq l}}^{Q(Q-1)/2} \left\{ \left[d^+(q_k, q_l) - d(q_k, q_l) \right] / d^+(q_k, q_l) \right\}^2$$

where $d(q_k, q_l)$ is the intercluster distance in the hypersphere and $d^+(q_k, q_l)$ is the intercluster distance in the representation plane; evidently, $d(q_k, q_l) > d^+(q_k, q_l)$. The minimization is made by a simplex method to obtained the percent. minimum standard deviation

$$SD(\%) = \left\{ E / \left[(Q(Q-1)/2) - 1 \right] \right\}^{1/2}$$

The clusters are then represented in the plane as circles centered in $\underline{p}^{(q)}$ with dispersion ϑ as a radius

$$\vartheta = arc\ sin \left[1 - \underline{c}^{(q)2} \right]^{1/2}$$

Along with this representation of the clusters, the representation of individual objects in the plane has also been described. The method has been used to map the fatty acid content of Italian olive oils (Forina and Armanino, 1982).

2.4.4 REPRESENTATION BY FACES

Van der Voet et al. (1984) have recently used a rather
different visualization approach for clustering of wines,
based on two-dimensional pictures of human faces charac-
terizing multidimensional data. In this method, each variable is represented as a characteristic feature of the
human face with the value proportional to the size or to
the form of certain part of the face (for instance, the
concentration of an element in wine can determine the
size of the eyes). The method evidently tries to exploit
one of the most advanced of human pattern recognition
abilities - the ability to recognize an individual face
in a "crowd" of faces and, more importantly, to detect
a similarity between some faces.

2.4.5 AUDIO-DISPLAY

A new interesting display method has been proposed by
Yeung (1980). In this audio-representation (display) of
multivariate analytical data each measurement is trans-
lated into an independent property of sound. An example
for 9-dimensional measurement space has been given, but
it should be possible to approach up to dimensionality
twenty. The trained ear can reliably determine each sound
in relation to the value in the original measurement.

It has been recently underlined (Derde et al., 1983)
that the first step in the analysis of a multivariate
data set should always be the display, as in this way an
indication can be obtained of what can be expected from
further multivariate investigations by clustering and/or
supervised methods.

Under the term DISCLOSE, Bawden (1984) has described an approach using integrated set of classification and display routines; a set of seven different multivariate display procedures included in the DISCLOSE provides reliable insights into the structure of complex data sets.

CHAPTER 3
Applications of Chemical Pattern Recognition

There exist many recent applications of chemical pattern recognition using various types of chemical data (Section 3.1) for different classification problems (Section 3.2). Chemical data are quite suitable for solving many real problems, ranging from physical ones up to very complex biological problems.

Chemical data are gathered and arranged in the enormous chemical literature, which is well organized. Nevertheless, it is a very cumbersome task for a user to extract information stored in these data compilations in a realistic time without the assistance of computers. Pattern recognition is one of the tools successfully tested in the modern computer-aided extraction of information contained in chemical data bases. Actually, recent results indicate that the fructification of the enormous and exponentially increasing amount of chemical data is possible in practice only by efficient processing using computers. Such a processing is highly desirable for the control of the future experiments to be optimized in a feedback way.

Thus, for example, the great importance of more rational (and thus economical) drug design on the basis of retrieved information is well known; pattern recognition is a prospective candidate for solving this problem (Subsection 3.2.3). The "pure" chemical problems deal with

classification of chemically characterized objects according to some aspect of their chemical behaviour (Sub-
section 3.2.2); for example, chemically described metals
can be classified with respect to their behaviour in
certain catalytic reactions. Finally, the physical behaviour can be interpreted using chemically represented objects (Sub-section 3.2.1) if an appropriate type of feature is selected.

Several practical applications are summarized in this
Chapter with the aim of demonstrating some recent trends
in the use of chemical pattern recognition.

3.1 CHEMICAL DATA

Raw data should be preprocessed before a pattern-recognition analysis for the elimination of non-inherent, artificial properties, e.g., for the elimination of the effect
of different scales (units). This is especially important
in the case of so termed "inhomogeneous" data sets which
can be, after an appropriate preprocessing, successfully
used in pattern recognition. This effect is eliminated by
scaling. The most commonly used is the so called autoscaling, i.e., normalization of each variable such that
its mean be zero and the standard deviation be unity

$$x'_{ij} = (x_{ij} - \overline{x}_i)/(\overline{x_i^2} - \overline{x}_i^2)^{1/2}$$

where x'_{ij} are autoscaled data, x_{ij} are original data,
$\overline{x}_i$ is the mean of the data for the i-th variable and $\overline{x_i^2}$
is the mean of squared data.

The normalization has to be used critically. For example,
the normalization by the range or the standard deviation
may yield distinct differences in the results, if the

ratio of range to standard deviation is not roughly constant for all variables (Bawden, 1983). In geochemistry
and geology the well-known closure (constant sum) problem
has been recently discussed by Johansson et al. (1984);
they found that a closure effect on the data (and consequently on the results) is also a frequent case in analytical chemistry, e.g. in chromatography and mass spectrometry, and that it should be avoided to minimize this
problem.

In many cases, a function of variable x_i

$$x_i^{''} = f(x_i)$$

is used for the reduction of the dynamic range of measured
values, e.g., by using the variable in its logarithmic
form. Nevertheless, when one is not sure about the kind
of data function to be used, then it is not easy to decide
whether original data or some function should be used.
In order to avoid a possible bias in the conclusions,
Massart et al. (1982b) have recently recommended that all
the calculations be carried out on original data, except
for scaling.

Weighting of variables is closely related to the classification and, therefore, it is outlined in connection with
feature selection (Section 2.3).

3.1.1 ANALYTICAL DATA

This Sub-section deals with some applications of pattern
recognition using elemental concentration data and composition of mixtures, which are valuable in various identification problems. In the former case, a classical example
is presented in Kowalski's archaeological artifact data

(Kowalski et al., 1972) used as a standard system for evaluation of performance of pattern recognition methods by direct comparison (e.g. Coomans and Massart, 1981a).

Recently, computer-aided chemometric methods including pattern recognition have been applied to the evaluation of major, minor or trace element data for very different purposes (Boulle and Peisach, 1979; Kwan et al., 1979; Kwan and Kowalski, 1980; Wegscheider and Leyden, 1979; Nauer and Haevernick, 1980; Massart, 1981a; Pillay and Peisach, 1981; Blasius et al., 1982; Massart et al., 1982b; Matsubara, 1982; Scarminio et al., 1982; Arunachalam et al., 1983; Bisani et al., 1983; Borszéki et al., 1983; Watterson et al., 1983; Arunachalam and Gangadharan, 1984; Carpenter and Till, 1984; Danzer et al., 1984; Hitchon and Filby, 1984; van der Voet et al., 1984).

The concentration values were estimated by means of different analytical methods:
- atomic absorption (Scarminio et al., 1982; Blasius et al., 1982),
- emission spectrometry (Kwan et al., 1979; Scarminio et al., 1982) and inductively coupled plasma-optical emission spectrometry (ICP-OES) (Carpenter and Till, 1984),
- neutron activation analysis (Boulle and Peisach, 1979; Watterson et al., 1983; Hitchon and Filby, 1984),
- proton-induced x-ray emission (PIXE) analysis (Boulle and Peisach, 1979; Pillay and Peisach, 1981),
- energy-dispersive X-ray fluorescence (EDXRF) spectrometry (Wegscheider and Leyden, 1979),
- X-ray fluorescence analytical flame photometry (Blasius et al., 1982) and
- laser microspectral analysis (Borszéki et al., 1983).

The processing of obtained signals to the concentration values is not a matter of discussion here because it is thoroughly treated in the specialized literature of analytical chemistry; the data for pattern recognition treatment are directly the respective concentration values. Elemental analytical data can be used, moreover, in combination with other types of data (Kwan and Kowalski, 1980; Matsubara, 1982). Evidently, pattern recognition analysis permits, in many cases, the extraction of non-measurable complex information from elemental patterns produced by very efficient modern analytical methods.

The second main type of analytical data suitable for identification purposes is the composition of mixtures. Several examples are given in the respective Sub-sections of this Section and the Section 3.2.

3.1.2 PHYSICO-CHEMICAL PARAMETERS OF ELEMENTS

Fundamental physico-chemical variables are highly suitable for representation of simple molecules, (see Section 1.2 and Sub-section 3.1.3) and for characterization of elements themselves with respect to some of their classification properties.

In the latter case, elemental physico-chemical parameters have been used in modelling catalytic activity of metals (see Section 1.2) in hydrogenolysis of ethane (Štrouf et al., 1981a) and in modelling the initial steps of the catalytic Fischer-Tropsch synthesis: chemisorption of hydrogen (Kuchynka et al., 1981) and chemisorption and dissociation of carbon monoxide (Štrouf et al., 1981b). The pattern recognition modelling yields a set of linearly independent variables, which were employed in a polynomial approximation of the respective classification property by simplex (Štrouf and Fusek, 1981) or by

regression optimization. The catalytic activity in hydrogenolysis of ethane was approximated by a polynomial of
second order (Kuchynka et al., 1981a) and the adsorption
heats of hydrogen (Štrouf et al., 1982) and carbon monoxide (Fusek et al., 1983) by polynomoals of first order.

Pijpers and Vertogen (1982) employed pattern recognition
techniques to find relevant physico-chemical parameters
of elements with respect to their superconductivity properties (see Section 1.2).

3.1.3 CHEMICAL STRUCTURE

Molecular structure represents a highly concise information source for data analysis methods, including those
dealing with classification. However, the crucial problem
is the encoding of the structure into an appropriate form
for computerized pattern recognition with minimum loss of
discriminatory information.

Formal structural data (numerical descriptors) were
frequently used for the pattern recognition purposes (see
e.g. Brugger et al., 1976). This rather formal approach
is highly suitable for processing of various types of
compounds (Section 1.1). Unfortunately, the results obtained from using these descriptors do not have physical
meaning and they cannot be used for modelling purposes.

For modelling (Section 1.2) it seems to be more suitable to adopt physically significant variables for elements (physico-chemical parameters as, e.g., electronegativity, heat of fusion, electrical conductivity)(see
preceding Sub-section). A serious limitation is the
applicability of this approach only for very simple compounds. For modelling of a substituent effect, physico-
chemical substituent parameters (constants) are commonly
used (for example, Hammett constants or the Hansch

lipophility parameter). In this case, the use is limited to a single structural type of compound, which differs only in the substituents.

The most sophisticated representation can be seen in calculated theoretical structural features. The analysis results using this type of variables can be discussed from a theoretical point of view. Thus, Miyashita et al. (1981) have recently represented polycyclic aromatic hydrocarbons, and their possible metabolites, by a variety of reactivity indices taken from simple Hückel molecular orbital theory, and then they have carried out a pattern recognition study of the carcinogenic process caused by these compounds. By quantum-chemical identification, steroid hormones have been represented for pattern recognition analysis of their biological activity by structural fragments (chromophores), further modification of which does not have substantial effect on the calculated electron transition parameters (Barenboim et al., 1982). Lukovits (1983) has studied quantitative structure-activity relationships employing independent quantum chemical indices as determined by principal component analysis. Theoretical variables derived from quantum chemical or other theoretical calculations have been proposed for modelling of a relationship between chemical structure and biological activity by the SIMCA pattern recognition (Dunn and Wold, 1980).

Generally, the basic requirement for proper encoding of chemical structure is a condition of representation of structural similarity in molecule. Randić and Wilkins (1979) have used a graph theoretical approach to recognize the molecular similarity. Some heuristics for KNN searching in chemical structure files has been described by Willett (1983). Correlation between chemical structure and some classification property was investigated by different hierarchical agglomerative clustering algorithms

(Willett, 1982a); the Ward algorithm (see Everitt, 1974) gave the best prediction results, whereas the KNN algorithm (see Sub-section 2.1.2) yielded the worst results.

Martin and Panas (1979) have discussed mathematical considerations in a drug series design and formulated the following characteristics for a series:

(a) the analogues should be synthetically feasible,
(b) the series should contain enough variation in the properties that may influence the searched potency,
(c) these properties should be varied independently of each other and
(d) the series should be of the minimum acceptable size.

Nevertheless, the "natural" training set of compounds available for drug design study is often rather incomplete and unbalanced and, therefore, it is not sufficiently representative for the study. Austel (1983) proposed a design of a representative learning series of structures by combined application of 2^n factorial schemes and pattern recognition techniques. Dove, Streich and Franke (Dove et al., 1980; Streich et al., 1980) have proposed a mapping approach for rational selection of substituents for test series of compounds.

Moreover, in structure-activity pattern recognition analysis the asymmetric case is relatively frequent when active compounds are distinguished from inactive ones (Dunn and Wold, 1980a). For this case, such disjoint methods as are the SIMCA or the SPHERE methods (Sub-section 2.1.3) could be useful.

It is worth mentioning that the characterization of chemical structures using information theoretic indices (Bonchev, 1983) could be an advantageous possibility to encode chemical compounds for data analysis in a very concise form.

Physico-chemical parameters of elements

In the case of relatively simple inorganic molecules, the
physico-chemical parameters of the elements (Sub-section
3.1.2) composing the molecule can be used as has been
recently shown for complex hydrides by Wold and Štrouf
(1979). For complex hydrides substituted by ligand(s),
combination of physico-chemical parameters for elements
with descriptors for ligand has been advantageously applied (Štrouf and Wold, 1977; Fusek and Štrouf, 1979,
1980). Also the combination of these parameters for elements with parameters for substituents (see below) can
be used for the representation of monosubstituted hydrides
(Wold and Štrouf, 1979a). The use of physico-chemical
constants has been proposed by Yuan C. et al. (1981) for
the representation of oxygen-containing phosphorus-based
ligands in a reactivity study.

Physico-chemical parameters of substituents

For a series of organic compounds belonging to a single
skeletal type (steroids, aromatics etc.), which are substituted by various substituents, the use of physico-
chemical parameters of the substituents is preferred to
other representations. The parameters are thoroughly and
critically studied in connection with linear free energy
relationships (Chapman N.B. and Shorter, 1972, 1978).
The substituent parameters (e.g. the Taft constants) have
been used for pattern recognition classification of
azomethines in relation to their photochromic properties
(Chemleva et al., 1982). It is worth mentioning that the
substituent parameters have shown a strong grouping and
a high collinearity in a recent multivariate analysis

(Alunni et al., 1983); therefore, they should be treated in pattern recognition analyses in a very careful way and, moreover, separate models for the classes are recommended.

In a large number of applications, the substituent parameters are used in structure-activity relationship studies as, e.g., for modelling a relationship between chemical structure and biological activity by the SIMCA pattern recognition (Dunn and Wold, 1980). The substituent constants form a basis of the representation of biologically active derivatives in many structure-activity studies (see e.g. Dove et al., 1979; Ogino et al., 1980).

Molecular structure descriptors

Molecular structure descriptors have been used in a series of papers (e.g. Adamson and Bawden, 1975, 1976, 1980; Chou and Jurs, 1979a; Jurs et al., 1979; Moriguchi et al., 1980; Adamson and Bawden, 1981; Moriguchi and Komatsu, 1981a; Yuta and Jurs, 1981; Dunn et al., 1982; Miyashita et al., 1982a; Rose and Jurs, 1982).

Recently, Bawden and Fisher (1985a) have described a novel measure of the effectiveness of screen set performance for chemical substructure searching; the measure is suitable for long-term monitoring of system performance and for comparison of screening performance in different searching systems. The substructural (fragment) descriptors are frequently employed in computer-aided assessment of structure-activity relationships (Henry and Block, 1979, 1980; Hodes, 1981, 1981a; Tinker, 1981, 1981a; Henry et al., 1982; Bodor et al., 1983, 1983a; Klopman, 1984; Solominova et al., 1984). The fragment molecular connectivity values can be extracted from an appropriate connection table (Henry and Block, 1979, 1980; Tinker,

1981a) and by using Wiswesser line notation (Adamson and Bawden, 1975, 1976). The intermolecular similarity can be calculated for molecules characterized by lists of substructural fragments (Willett, 1982).

Also some other papers are concerned with chemical structure-biological activity problems solved by pattern recognition (Takahashi et al., 1980, 1980a; Yuan M. and Jurs, 1980; Dunn and Wold, 1981, 1981a; Mager, 1981; Miyashita et al., 1981a, 1981b; Moriguchi et al., 1981; Jurs, 1983, 1983a; Peredunova and Kruglyak, 1983; Wold et al., 1983a; Stouch and Jurs, 1985a; Jurs et al., 1985). These problems are included in books (e.g. Stuper et al., 1979) and in reviews (Miyashita et al., 1981c, 1982; Whalen-Pedersen and Jurs, 1981; Jurs et al., 1983b). Wold and co-workers (1982) have published a critical survey on applications of pattern recognition to structure -activity problems and discussed conditions for the applicability of multivariate structure-activity relationships (Wold and Dunn, 1983).

A similar problem is the correlation between chemical structure and sensory property; for example, Takahashi et al. (1982) have investigated the relation between structure of several perillartine derivatives and their taste quality. Jurs et al. (1981) have discussed computer-assisted studies of chemical structure and olfactory quality.

An effort has been made to extract information about structural similarities of molecules by graphical and numerical techniques yielding visualization of linear relationships found between molecules or fragments (Watkin, 1980).

3.1.4 SPECTRAL DATA

Clerk and Székely (1984) have recently suggested that
today's pattern recognition approaches to a spectra-
structure study cannot work successfully due to the lack
of suitable models. The merits and demerits of data retrieval and artificial intelligence systems for identifying polyatomic molecules from their molecular spectra
have been considered by Gribov (1980, 1984). The success
of retrieval and pattern recognition systems in spectroscopic analysis depends on the quality of the used spectra
library and of the training set, respectively (Kwiatkowski
and Riepe, 1984).

Mass spectra

Analysis of low resolution mass spectra is the most frequent application of pattern recognition in the spectral
data field (Chapman J.R., 1978; Varmuza, 1980, 1984b);
more than 100 papers have been published within 1969-1981
(Varmuza, 1984b). Evidently, the suitable form of mass
spectra seems to be one of the reasons of a popularity of
these spectra in pattern recognition analysis. Digitization
is not required by these spectra: simply the preprocessed
peaks at the mass numbers can be taken as features in a
pattern formation. Nevertheless, in high-precision chemical structure classification, autocorrelation transforms
should be employed to counter-act the effect of mass spectral shifts (Wold and Christie, 1984b).
  It should be mentioned here that many new chemical
pattern recognition methods have been tested using just
mass spectral data. For instance, measure of confidence
based upon a weighted average of a set of backward probabilities has been proposed for learning machine (Sub-

section 2.1.1) trained for multicategory classification of mass spectra (Richards and Griffiths, 1979).

Recent applications of pattern recognition for mass spectra analysis are included in books (e.g., Bailley, 1979; Mellon, 1979; Varmuza, 1984b) and reviews (McLafferty and Venkataragharan, 1979; Meisel et al., 1979; Jellum, 1981a; Martinsen, 1981; Varmuza, 1983; Ten Noever et al., 1984; Varmuza and Lohninger, 1984).

Mass-spectral peaks are sometimes preprocessed to binary or logarithmic forms (e.g. Maclagan and Mitchel, 1980). More discriminatively effective features can be derived from mass spectra than are "native" simple peak heights (Anderegg, 1981). Recently Varmuza and Lohninger (1984) have developed a new approach based on heuristic generation of features for each class separately using some idea of chemist's interpretation way.

As demonstrated by Domokos and Frank (1981), orthogonal transformations (viz. Fourier, Walsh and Haar transformations) can be useful for feature extraction from mass spectra. Principal component analysis was applied to the resolution of mass spectral data (Arunachalam and Gangadharan, 1984) (see Sub-section 2.3.1); the procedure determines relative amounts of pure components in the mixtures with and without the use of pure mass lines. The usefulness of factor analysis of mass spectra was evaluated on simulated Gaussian gas chromatographic-mass spectrometric data with the objective to detect the number of components (Woodruff et al., 1981a).

Cluster analysis of mass-spectrometric data was used to simplify the computerized identification of compounds in large spectral collections; first, large groups of compounds with similar mass spectra characteristics are replaced by a suitable representative pattern and, in the second step, the representatives are investigated instead of all spectra (Domokos et al., 1980).

Mass-spectral library search can be improved by inves-
tigating sequential spectra in series of spectra; masses
with highly correlating sequential intensities are clustered into individual groups which belong to the same
component spectrum (Domokos and Henneberg, 1984a). Moreover, identification of pure compounds and mixtures by
comparison of mass spectra can be carried out by a new
identity-orientated search procedure for mass spectral
libraries named IDS (Domokos et al., 1984). Also a combination of forward and reverse searching procedures can be
used for unknown mass spectra of both pure compounds and
mixtures without prior knowledge of purity (Stauffer et
al., 1985).

Mass spectra-structure correlations were evaluated by
a training algorithm including a pattern recognition technique and an artificial intelligence program (Jerkovich,
1980). Mass spectra of nucleosides were interpreted by
four pattern recognition techniques (distance from the
mean, the learning machine, statistical linear discriminant analysis and the KNN method) (see Section 2.1) to
search for a set of various structural characteristics
(Maclagan and Mitchel, 1980). For structural purposes,
a correlation coefficient pattern recognition technique
has been used (Mahle and Ashley, 1979). An interpretation
of mass spectra of steroids has been published using binary classifiers with a continuous response (Varmuza and
Rotter, 1980b).

Field-desorption and fast atom-bombardment mass spectrometric profiles of complex mixtures of non-volatile compounds (e.g. components of urine or wine samples) were
evaluated by Fisher weighting (Sub-section 2.3.2) and
principal component analysis (Sub-section 2.3.1) followed
by classification by KNN method (Sub-section 2.1.2) and
nonlinear mapping (Sub-section 2.4.3) (van der Greef et
al., 1983).

The modern gas chromatography-mass spectrometry analyses produce large data files. For their interpretation a computer assistance is necessary. Thus, pattern recognition techniques have been recently used for identification of pentafluoropropionyl dipeptide methyl esters (Ziemer et al., 1979), gasolines (Bertsch et al., 1981) and polycyclic aromatic hydrocarbons (Varmuza and Lohninger, 1984).

Highly complex data produced in pyrolysis-mass spectrometry can be also analysed by pattern recognition methods. Meuzelaar (1982) characterized coals and coal liquids and Tsao and Vorhees (1984) did the same with smoke aerosols from nonflaming combustion by pattern recognition analysis of the respective pyrolysis-mass spectrometric profiles. Particularly interesting is pattern recognition analysis of pyrolysis-mass spectral data of complex biological samples (e.g., Windig et al., 1983; Boon et al., 1984). Most recently, Garozzo and Montaudo (1985) have described identification of polymers by pattern recognition analysis and library search of mass spectra obtained by direct pyrolysis-electron impact mass spectrometry.

Infrared spectra

Information on digitized infrared spectra can be used for identification as well as for classification purposes. In the former case, pattern recognition methods can be applied, e.g., to an analysis of the library of Fourier transformed infrared spectra which is not error free and, moreover, it can contain spectra of complex mixtures (Frankel, 1984). Zupan (1982, 1982a) has published a new fractal (3-distance) clustering method (Sub-section 2.2.1) suitable for the library of Fourier transform infrared spectra. The computer-aided interpretation of vibrational spectra taken from infrared and Raman files with respect

to structural units has been reported (Visser and van der
Maas, 1981). Recently, Delaney et al. (1985) have described three distinct approaches applicable for optimization of a similarity metric for matching highly compressed vapour-phase infrared spectra.

Classification of dithiocarbamate complexes on the basis
of infrared spectroscopic data combined with ^{13}C-NMR data
has been made by means of unsupervised and supervised
techniques of pattern recognition (Pijpers et al., 1979).
The principal component technique using multiple linear
regression optimized by an equivocation criterion was
applied for classification of organic compounds with
respect to the presence of a tert-butyl group in their
molecule (Bink and van't Klooster, 1983). Monosubstituted
phenyl rings have been classified by a statistical linear
discriminant method (Tsao and Switzer, 1982; Tsao, 1982b)
and by branching tree methods (Tsao and Switzer, 1982a).

On-line infrared spectral data of organic compounds can
be analysed by a hierarchically organized computerized
system ASSIGNER to give the probability of the presence
of functional groups (Farkas et al., 1980). Another computerized infrared spectral interpreter based on binary
decision trees has been also described and tested on the
aldehyde functionality (Woodruff and Smith, 1981). Zupan
et al. (1980a) have described an interactive procedure
for the analysis of infrared spectra of mixtures implemented on a minicomputer.

A remarkable interest has been recently focused on
pattern recognition analysis of vapour-phase infrared
spectra as a detection method in gas chromatography. Thus,
Delaney and his co-workers (1979) tested the suitability
of pattern recognition methods (hyperplane separation
methods and KNN methods, see Sub-sections 2.1.1 and 2.1.2,
respectively) for distinguishing functional groups and
the presence of specific elements in compounds from their

vapour phase infrared spectra. The classification of
organic compounds according to the presence or the absence
of certain selected functional groups was performed by
Domokos et al. (1983) by various unsupervised as well as
supervised pattern recognition methods. For some simple
gaseous molecules, digitized binary vectors were extracted
from rotational fine structure infrared spectral data and
used for pattern recognition analysis of a mixture of
these gases in the presence of water (Honeybourne and
Smith, 1982). The on-line identification of gas chromatographic peaks can be based on interferometric data
from gas chromatography-Fourier transform infrared data;
the classification of compounds with respect to the presence of functional groups has been made by means of a
pattern recognition technique, e.g., by linear learning
machine (see Sub-section 2.1.1) (Hohne et al., 1981).

The Isenhour group has recently studied the approaches
to the reconstruction of gas-chromatography Fourier-trans-
form infrared chromatograms (Hanna et al., 1979, 1979a;
Hohne et al., 1981; Wieboldt et al., 1980; Sparks et al.,
1982; Owens et al., 1982). Most recently, Isenhour with
his co-workers has described methods for reconstructing
chromatographic data from liquid chromatography-Fourier
transform infrared spectrometry (Wang et al., 1984).

Nuclear magnetic resonance spectra

Recent applications of pattern recognition for the interpretation of NMR data are not numerous. Thus, e.g., the
NMR data of a series of 4-substituted styrenes have been
interpreted by the SIMCA method (Edlund and Wold, 1980).
NMR data (together with other, e.g., elemental and vapour
osmotic, data) were used to characterize heavy oils using
statistical methods, namely pattern similarity analysis

and principal component analysis (Matsubara, 1982). In
a computerized analytical system called ASSIGNER, on-line
measured spectral data (including ^{13}C and ^{1}H data) are
compared with off-line collected parameters in search for
functional groups in organic compounds (Farkas et al.,
1980).

For ^{13}C-NMR spectra, substituent effects have been
studied by principal component analysis for chalcones and
their analogues (Musumarra et al., 1981), α-phenyl-β-
(2-thienyl)acrylonitrile derivatives (Musumarra et al.,
1981a), triazenes (Dunn et al., 1982a) and monosubstituted
benzenes (Johnels et al., 1983a). This analysis has been
applied also to feature extraction from ^{13}C-NMR spectra
(Arunachalam and Gangadharan, 1984). The prediction of
^{13}C-NMR chemical shifts using partial least-squares (PLS)
data analysis (see Sub-section 2.1.3) has been recently
described (Johnels et al., 1983). Simplified ^{13}C-NMR parameters of polycyclic aromatic hydrocarbons have been related to their carcinogenic potency by means of the PLS-
method (Nordén et al., 1983). With the aid of pattern
recognition techniques of the ARTHUR package (Harper et
al., 1977), the dithiocarbamates have been divided into
five chemical classes on the basis of ^{13}C-NMR chemical
shifts of the carbon atom of the NCS_2 moiety (van Gaal et
al., 1979).

The ^{15}N-NMR chemical-shift data for triazenes have been
analysed by principal component analysis (Dunn et al.,
1982a).

Pattern recognition in one-dimensional spectra can be
misleading, since a multiplet cannot be distinguished
a priori from an accidental juxtaposition of chemical
shifts; on the other hand, two-dimensional NMR spectra
provide information sufficient to avoid pitfalls in pattern recognition analysis (Meier et al., 1984).

Other spectra

Pattern recognition and factor analysis were applied to
Fourier-transformed X-ray-excited carbon Auger spectra to
extract chemical information from line shapes
(Gaarenstroom, 1979).
A nine-dimensional vector was used for the characterization of X-ray valence band spectra of a set of copper
compounds and transformed to a 2-dimensional representation (Drack et al., 1979).
Low-temperature luminescence and room-temperature
fluorescence spectral data of polynuclear aromatic hydrocarbons were assessed by three different measures of
similarity to evaluate the variability in the spectra
measured under different instrumental conditions (Sogliero
et al., 1982).
Severely overlapping fluorescence spectra of a mixture
of humic acid, ligninsulfonate and an optical whitener
from a detergent were treated by the PLS method to
quantitatively determine the individual components
(Lindberg et al., 1983; Sjöström et al., 1983).

3.1.5 ELECTROCHEMICAL DATA

Electrochemical data represent a valuable source of information which can be successfully extracted by pattern recognition methods. Therefore, this data analysis attracted
a high interest also in the past period. This can be demonstrated by a series of about twenty works published
from 1979 to 1985 mainly by the Perone's and Ichise's
groups. Recent progress in deviation-pattern recognition
interpretation of electrochemical data using computerized
pattern recognition has been reviewed by Rusling (1984).
Generation of an electrochemical data base for pattern

recognition has been studied by Byers and Perone (1983).
A series of papers describe the use of pattern recognition
techniques in qualitative electrochemical analysis (Byers,
1981; Byers et al., 1983a; Ichise et al., 1980, 1980b,
1982, 1983).

For pattern recognition applications in quantitative
electrochemical analysis, learning machines (Bos, 1979;
Ichise et al., 1980a) or deviation-pattern recognition
(Meites, 1982) have been used. The latter technique has
been employed for mechanistic classification of one-
electron potentiostatic current-potential curves (Rusling,
1983, 1983a).

The Perone group has also studied the classification of
voltammetric data by pattern recognition (DePalma and
Perone, 1979, 1979a; Schachterle, 1980; Schachterle and
Perone, 1981).

Particularly interesting is an application of pattern
recognition to lifetime prediction of batteries from
initial cycling test data as described by Perone and his
co-workers: the concept of specific lifetime prediction
was first explored for sealed Ni/Cd space cells and the
results of this study demonstrated that the cells from the
same production lot, with similar fabrication and operational conditions, could be categorized as "short-lived"
and "long-lived" cells from initial measurements made on
them with virtual 100% accuracy (Byers and Perone, 1979).
Similarly lead-acid batteries with about 87% accuracy
were classified (Perone and Spindler, 1984).

3.1.6 CHROMATOGRAPHIC DATA

Recently, the applications of pattern recognition methods
in chromatography have been reviewed (Isaszegi-Vass et al.,
1984).

Gas-liquid chromatography

In gas-liquid chromatography, characterization and selection of stationary phases can be made by principal component (factor) analysis (see Sub-section 2.3.1) (Fellous et al., 1981, 1982, 1983) as well as by hierarchical clustering (Sub-section 2.2.1) and minimal spanning tree (Sub-section 2.2.2) techniques (Huber and Reich, 1984). Clustering (numerical taxonomy) of common phases has been tested using chloroperoxy alkyl esters (De Beer and Heyndvickx, 1982).

Pattern recognition can be useful also in the deconvulation of chromatographic data (McCown et al., 1982). The reliability of the identification of individual compounds by chromatography can be improved by pattern recognition analysis using calculated and reference retention data (Stepanenko, 1982). Gas chromatographic retention data can be analysed by a pattern recognition method to extract information on the chemical structure of monofunctional compounds (Huber and Reich, 1980).

Pattern recognition has been used to advantage in gas chromatography for the identification of compounds in various complex mixtures such as crude oils (Clark and Jurs, 1979), wines (Kwan and Kowalski, 1980a), essential oils (Chien, 1985) and polychlorinated biphenyls (Lea et al., 1983; Dunn et al., 1984).

Particularly interesting is an application of gas chromatography-pattern recognition to sensory evaluation of aroma in soy (Aishima et al., 1979) and to comparison of the cuticular hydrocarbon patterns between different colonies of fire ants (Brill et al., 1985, 1985a).

Gas chromatographic analysis of multi-component mixtures by pattern recognition technique can be implemented by using a microcomputer (Saarinen, 1983).

Capillary-gas chromatography provides a very complex profile, from which information can be extracted by pattern recognition as has been shown for volatile constituents of various biological samples: human urine (McConnell et al., 1979), serums (Zlatkis et al., 1979) and brain tissues (Jellum et al., 1981; Wold et al., 1981), as well as muscle and gonad tissues of blue mussels (Kvalheim et al., 1983). The research group at Philip Morris has studied the use of pattern recognition analysis of capillary-gas chromatograms of cigarette smoke (Parrish et al., 1981, 1983; Hsu et al., 1982). Capillary-gas chromatography and HPLC (see below) of volatile plant substances have been used to determine pest and disease resistance in the breeding (Lundgren et al., 1981).

Headspace gas chromatograms have been used to distiguish between nontoxic and toxic (botulinum-positive) fish samples (Snygg et al., 1979).

Similarly, pyrolysis-gas chromatography yields very complex chromatograms; therefore, multivariate analysis methods have to be adopted for the interpretation of these chromatograms. The reproducibility of repetitive pyrolysis -chromatograms has been examined (Blomquist et al., 1979b) and these chromatograms were used for classification of species (Blomquist et al., 1979) and strains (Blomquist et al., 1979a) of Penicillium. Pino et al. (1985) have applied pyrolysis-gas chromatography-pattern recognition to the detection of carriers of the cystic fibrosis gene. The multivariate analysis of pyrolysis-chromatograms was also applied to other types of mixtures from pyrolysis (Klee et al., 1981; Morgan, 1981; Söderström et al., 1982; Milina et al., 1983).

Most recently, chemometric approaches to gas chromatography have been reviewed (Harper, 1985).

Liquid chromatography

Classification of coal liquids by reverse-phase high-
performance liquid chromatography (HPLC) was performed by
pattern recognition interpretation of curves from three
different detectors (Sepaniak and Yeung, 1981). Different
pattern recognition techniques were used to investigate
the intrinsic structure of HPLC profiles obtained from
plasma (Scoble et al., 1983). Chromatograms of urinary
proteins obtained in a fast protein liquid chromatography
system have been examined and classified by cluster analysis (Marshall et al., 1984).

Thin-layer chromatography

Some structural information about amines and related compounds can be extracted from R_f thin-layer chromatographic
data (Carbone et al., 1983).

3.2 CLASSIFICATION PROPERTY

3.2.1 PHYSICAL PROPERTY

Recently, attempts to model a physical classification
property on the basis of physico-chemical parameters have
been carried out. Thus, various chemical-bond relating
properties of inorganic compounds can be predicted using
parameters such as, e.g., pseudopotentials and Slater's
atomic orbitals; specifically, crystal types of inter-
metallic compounds and thermodynamic behaviour of oxide
systems have been studied (Chen N.Y. et al., 1981).
Also the superconductivity behaviour of chemical

elements can be successfully modelled by pattern recognition techniques (Pijpers and Vertogen, 1982) (see Section 1.2).

Models predicting alloying behaviour of two-element systems have been critically assessed by the use of pattern recognition techniques (Kentgens et al., 1983). The prediction of electrical conductivity of alloys can be also made (Belozerskikh and Dobrotvorskii, 1980) as an application of these methods.

Chemleva et al. (1982) have modelled photochromic properties of azomethines by pattern recognition techniques.

3.2.2 CHEMICAL PROPERTY

The stability of complex hydrides has been modelled by Wold, Štrouf and Fusek (Wold and Štrouf, 1979, 1979a; Fusek and Štrouf, 1979) by adopting sophisticated pattern recognition methods (see Section 1.2 and Sub-section 2.1.3).

The reactivity of compounds containing various phosphorus-based ligands in extraction of rare earth elements has been studied by Chinese authors (Yuan C. et al., 1981). Ryabova et al. (1983) have used pattern recognition in the elucidation of the effect of the most important factors in the in situ epoxidation of ethylene-propylene rubber by performic acid. By means of a pattern recognition technique, van Gaal with his co-workers (1979) has divided the dithiocarbamates according to their ^{13}C-NMR shifts of NCS_2-fragment into five chemical classes.

Applications of pattern recognition in prediction and analysis of the mechanism of the action of heterogeneous catalysts have been reviewed by Ioffe and his co-workers (1980, 1983). Catalytic hydrogenation of fats and synthesis of thiophene were analysed by means of pattern recognition

from the point of view of selection of perspective catalysts (Amirova et al., 1980). A pattern recognition approach was proposed for formal evaluation of the classification of catalytic oxidation reactions with molecular
oxygen (Amirova and Masagutov, 1982). Also prediction of
the activity of polymetallic catalysts in hydrogen per-
oxide decomposition reaction has been studied with pattern recognition (Belozerskikh and Dobrotvorskii, 1980).
By means of the SPHERE pattern recognition technique,
modelling of catalytic activity of transition metals in
hydrogenolysis of ethane has been performed (Štrouf et al.,
1981a). Similarly, modelling of the catalytic Fischer-
Tropsch synthesis has been published (Kuchynka et al.,
1981; Štrouf et al., 1981b) (see Section 1.2).

In electrochemistry, pattern recognition has been
applied for qualitative (Ichise et al., 1983) and quantitative (Ichise et al., 1980a) simultaneous determination
of some metal ions, especially of Pb^{2+} and Tl^{+} ions.
Moreover, pattern recognition methods have been used for
mechanistic classification of electrode processes
(DePalma and Perone, 1979a; Schachterle and Perone, 1981;
Ichise et al., 1982; Rusling, 1983, 1983a).

One of the most important goals of pattern recognition
is the identification and/or the determination of components in different complex mixtures using data interpret-
able by the analyst only in a difficult way. For these
purposes, pattern recognition has been used, e.g., to
interpret gas and liquid chromatographic data (Lea et al.,
1983), spectrophotometric data (Lindberg et al., 1983)
and mass spectral data (Varmuza and Rotter, 1980b;
Woodruff et al., 1981; Arunachalam and Gangadharan, 1984).

The detection of bases in three-component mixtures can
be achieved by a pattern recognition analysis of potentiometric titration curves (Bos, 1979). The detection of
a minor weak acid in another weak acid can be made by an

analysis based on calculated potentiometric acid-base
titrimetric curves using deviation-pattern recognition
(Meites, 1982). This method is also useful in detection
of a minor reactant present in a pseudo-first-order
reaction system (Meites, 1982a).

An interesting application of pattern recognition is
the determination of carbon in molten steel by analysing
plateaus on thermograms (Fainzil'berg, 1980).

Another important objective of pattern recognition in
chemistry is to elucidate chemical structure from different types of data as has been mentioned in the respective
Sub-sections of Section 3.1. Here, only some examples are
summarized. Thus, a series of papers deal with the classification (identification, differentiation, distinguishing)
of organic compounds according to the presence (or the
absence) of functional groups. For these purposes,
chromatographic data (Delaney et al., 1979; Huber and
Reich, 1980; Carbone et al., 1983) and spectral data
(Mahle and Ashley, 1979; Farkas et al., 1981; Hohne et
al., 1981; Woodruff and Smith, 1981; Tsao and Switzer,
1982, 1982a; Domokos et al., 1983) have been mostly used.

Moreover, pattern recognition can be used in structural
characterization in some special problems as, e.g., in
the identification of amino acid sequences in poly-
peptides (Ziemer et al., 1979), in the classification of
structural types of nucleosides (Maclagan and Mitchell,
1980), N,N-dialkyl dithiocarbamate complexes (Pijpers et
al., 1979) and of benzenoic compounds (Adamson and
Bawden, 1981), in the prediction of the length of the
alkyl chain in poly(alkyl methacrylate) (Gaarenstroom,
1979), in the identification of isomers of tetrachlorodibenzo-p-dioxin (Nestrick et al., 1980), in distin-
quishing types of electrochemically reducible organic
compounds (Byers et al., 1983a) and in the investigation
of the secondary structure of procaryotic 5S-ribonucleic

acids (De Jong and Pijpers, 1985).

These examples demonstrate that a pattern recognition
approach can be useful in solving some special problems
in structural characterization. The approach often resembles that made by a human interpreter. The capability
of pattern recognition to identify structural similarities
on the basis of different features (or of their mixture)
is used with advantage in sophisticated artificial intelligence systems for processing data files (most frequently spectra libraries). However, for more fundamental
(causal) data-structure relationships, suitable physico-
chemical models are missing as has been discussed in the
case of spectral data-chemical structure relatioships
(Clerk and Székely, 1984) (compare Section 1.2).

Nevertheless, pattern recognition analysis including
data processing provides at least an insight into data
structure and suitability of these data for chemical
structure elucidation (step a), one of the main goals of
chemistry. After estimation and confirmation of the structure found, the analysis can search for a possible relation between the chemical structure and the classification
property in question (step b).

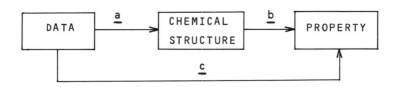

The analysis of data-property relationships is a more
"direct" way (step c), but extremely critical approach
to its use is necessary.

3.2.3 BIOLOGICAL PROPERTY

The use of pattern recognition in the classification of
chemical objects with respect to a biological property
is very popular (and also a criticized) application of
this multivariate data analysis.

Recently, the pattern recognition approach to biological activity of chemical compounds has become a common
part of books on drug design (Hodes, 1979; Jurs et al.,
1979a; Kirschner and Kowalski, 1979; Petit et al., 1979;
Stuper et al., 1979; Golander and Rozenblit, 1980; Lewi,
1980) and it has been reviewed (Miyashima et al., 1981c,
1982; Whalen-Pedersen and Jurs, 1981; Dunn and Wold,
1983; Ordukhnyan et al., 1983; Jurs, 1983a; Jurs et al.,
1985). The structure-activity problems have been thoroughly and critically treated by Wold, Dunn and co-
workers; they have discussed conditions for applicability
of multivariate quantitative analytic data methods including the pattern recognition methods and the factor
and principal component analyses (Wold and Dunn, 1983).
An assessment of structure-activity studies has shown
that about a half of the results are statistically inva-
lid (Wold et al., 1982). Dunn and Wold (1980a) showed
that in pattern recognition dichotomization of compounds
as active or inactive, the asymmetric case (i.e. the
active class is only structured) is very frequent and
only disjoint methods can be used in such cases. The
authors and their co-workers (Dunn et al., 1982) have
reported an application of partial least squares (PLS)
method for quantitative structure-activity relationships
between data from a battery of biological tests and a set
of chemical descriptors.

From the general point of view, the possibility of
employing very different types of data for estimation of
biological property is an extraordinary advantage. For

example, such extremely different types of data as fragment molecular connectivity values (Henry and Block, 1979,
1980) on one side, and quantum chemical indices (Barenboim
et al., 1982; Lukovits, 1983) at the other extreme have
been discussed in connection with pattern recognition.

On the other hand, recent studies of structure-activity
relationships have stimulated methodological progress of
pattern recognition. For instance, the development of the
SIMCA method seems to be catalyzed by problems in biological studies. Takahashi et al. (1980a) have applied the
simplex optimization of coefficients of a linear discriminant function in structure-activity studies.

Wold with his co-workers (Wold et al., 1983a) discussed
the formulation and applicability of semi-empirical and
empirical models relating chemical structure to biological activity, namely for toxicity. Prediction of acute
toxicity based on structural features of organic compounds
has been also reported (Solominova et al., 1984).

A very exciting problem at present is to predict a
carcinogenic property of compounds existing naturally and
artificially in our environment (Jurs et al., 1983b). The
study of the problem is extraordinary extensive and thus,
naturally, very expensive; suitable data-analytical
methods should be therefore used to assist in the objective evaluation of the obtained data bases, to extract
maximum value from the data. Pattern recognition is highly
convenient for its capability to analyse structure-
carcinogenicity relations even for chemical structures
represented by a heterogeneous data set (a set of data
composed of various types of features) (Jurs et al., 1979;
Yuta and Jurs, 1981). The main attention has been recently
focused on the carcinogenicity of two types of organic
compounds:

(a)   Dunn and Wold (1980), Yuta and Jurs (1980), Miyashita
       et al. (1981, 1982a), Whalen-Pedersen and Jurs

(1981), Nordén et al. (1983) and Klopman (1984) have
tested pattern recognition methods in structure-
carcinogenicity studies for aromatic hydrocarbons
and

(b)   Jurs et al. (Chou and Jurs, 1979; Rose and Jurs,
1982; Jurs, 1983), Dunn and Wold (1981, 1981a) and
Klopman (1984) for N-nitroso compounds.

Also the carcinogenicity of 4-nitroquinoline-1-oxides
has been studied by the SIMCA method (Dunn and Wold,
1980).

Tinker (1981, 1981a) has employed pattern recognition
techniques to relate bacterial mutagenesis activity to
chemical structure. Stouch and Jurs (1985b) have studied
molecular structure and genotoxic activity using a pattern
recognition technique.

Recent attempts to find a relation between chemical
structure and antitumour activity using a pattern recognition approach are primarily interesting from the point
of a more economical antitumour drug design. With respect
to antitumour potential, structural data were analysed
for mitomycin derivatives (Moriguchi et al., 1981),
withaferin analogues (Morigushi and Komatsu, 1981a) and
9-anilinoacridine (Henry et al., 1982; Jurs, 1983).

Particularly interesting are studies aiming to extract
information from tens of thousands of compounds of diverse
structural types as tested in the extensive National
Cancer Institute (the U.S.A.) prescreen program (Hodes,
1981, 1981a).

Moreover, other biological properties have been reported to be related to specific structural features,
e.g.:
- cardiotonic activity of 2-phenylimidazo[4,5-b]pyridines
  (Austel, 1983),
- anti-inflammatory activity of steroids (Bodor et al.,
  1983, 1983a) and phenylacetic acids and aminouracils

(Ogino et al., 1980),

- morphinomimetic activity of N-4-substituted 1-(2-aryl-
ethyl)-4-piperidinyl-N-phenylpropanamides (Mager, 1981),
- antibacterial activity of cephalosporins (Takahashi et
al.,1980; Miyashita et al., 1981a, 1981b) and other
types of antibiotics (Takahashi et al., 1980),
- hypotensive activity of N-alkyl-N''-cyano-N'-pyridyl-
guanidines (Morigushi et al., 1980),
- antiulcerous activity of benzoquanamines (Ogino et al.,
1980),
- central nervous activity of benzodiazepine derivatives
(Miyashita et al., 1981a; Peredunova and Kruglyak,
1983) and
- activity of β-adrenergic compounds (Dunn and Wold, 1980).

The Japanese group (Moriguchi et al., 1981) have applied
pattern recognition for assignment of a pharmacological
category to a set of diarylmethane-derived drugs.

Some attention has been also paid to special questions
of interaction between chemical compounds and biological
systems; thus adverse reactions of various drugs and
binding affinities of steroids for receptors have been
analysed (Morigushi et al., 1981). A structure-taste
(sweet or bitter) correlation of perillartine derivatives
using pattern recognition techniques has been also considered (Takahashi et al., 1982). The question of chemical
structure-olfactory quality relationships has been studied
by Jurs et al. (1981). The recognition of molecular sub-
structures responsible for the pesticidal activity of some
ketoxime carbamates can be carried out by a new program
introduced by Klopman (1984).

Pattern recognition methods, namely discriminant analysis and cluster analysis, have been found to facilitate
the formulation of a satisfactory structure-activity
correlation equation as has been shown by Chen B.-K.
et al. (1979) in a QSAR study of inhibition of dihydro-

folate reductase and thymidylate synthetase by quinazolines.

For structure-activity studies, the program package
ADAPT (Automated Data Analysis by Pattern recognition
Techniques) was developed (Stuper and Jurs, 1976; see
Stuper et al., 1977, 1979). ADAPT represents an interactive computer software system which consists of nine
main segments including, e.g., a file generator and a
descriptor developer, a preprocessor, feature selector
and a discriminant developer. The ADAPT method has been
recently used for a chemical structure-biological activity
relationships analysis as, e.g., for the relationships
between the structure of N-nitroso compounds and their
carcinogenic potential (Rose and Jurs, 1982).

Comparison and evaluation of searching techniques for
the retrieval of chemical toxicology information have
been discussed by Bawden and Brock (1985) for online
data-bases.

3.2.4 OTHER APPLICATIONS

Analytical chemistry

Recently, the pattern recognition approach has been
applied to objectively assess routine analytical methods
in clinical laboratories (Jansen et al., 1981, 1981a);
on the basis of data collected in the Netherlands National Coupled External/Internal Quality Control Programme
it has been found that analytical methods for determination of blood components can be described by the following features: (1) accuracy, (2) day-to-day precision,
(3) tendency to give erroneous results and (4) tendency
to give systematic errors in different laboratories.

Identification of materials
_____

Many papers deal with practical identification problems
as, e.g., with identification of:
- origin regions of Italian olive oils on the basis of
  fatty acids content (Derde et al., 1982a; Forina and
  Armanino, 1982; Forina and Tiscornia, 1982a),
- essential oils of different sources and in perfumes from
  gas chromatographic profiles (Chien, 1985),
- origin of wines based on organic compounds (Kwan and
  Kowalski, 1980a), amino acids (Derde et al., 1983) and
  elemental (Kwan et al., 1979) concentration data,
- quality of wines according to their elemental and organic compounds compositions in order to correlate
  objective measurements with subjective sensory evaluations (Kwan and Kowalski, 1980),
- wine samples from mass spectral profiles (van der Greef
  et al., 1983),
- Bordeaux and Bourgogne wines according to their chemical analysis (van der Voet et al., 1984),
- origin of milk samples (Coomans et al., 1981b, 1982c),
- crude oils by gas chromatography (Clark and Jurs, 1979),
- heavy petroleums on the basis of chemical analysis (e.g.
  elemental analysis, NMR spectra and vapour-phase
  osmosis) in respect to their behaviour during refining
  (Matsubara, 1982),
- asphaltenes in coal liquids from coal liquefaction
  processes by HPLC (Sepaniak and Yeung, 1981),
- brasses for forensic purposes according to their elemental composition (Carpenter and Till, 1984),
- iron meteorites on the basis of their physico-chemical
  data (Esbensen et al., 1984) and elemental composition
  (Massart et al., 1981),
- polymeric source materials from pyrolysis-mass spectra

(Tsao and Voorhees, 1984),

- cigarette types by gas chromatography of cigarette smoke (Parrish et al., 1981, 1983; Hsu et al., 1982),
- fungi by pyrolysis-gas chromatography (Blomquist et al., 1979a, 1979b) or by pyrolysis-mass spectrometry (Windig et al., 1983),
- bacteria by gas chromatography-mass spectrometry (Engman et al., 1984),
- Bacteroides gingivalis cultures by pyrolysis-mass spectrometry (Boon et al., 1984) and
- different colonies of imported fire ants (Solenopsis invicta, S. Richteri) by gas chromatography of their cuticular hydrocarbons (Brill et al., 1985, 1985a).

A very original approach to material characterization is the use of pattern recognition methods in the evaluation of information that may be gained from acoustic emissions produced by polymers under stress (Betteridge et al., 1980). The sound produced by a sample of a stressed polymer is evaluated as a function of the stress with two objectives:

(a) to predict mechanical behaviour of the tested polymeric material and

(b) to search for molecular events (bond breaking) causing the acoustic emissions.

Finally, it is worth mentioning that pattern recognition and cluster analysis methods can be used to assess food quality on the basis of subjective sensory evaluations received by skilled panel members and consumers (Molnár et al., 1984).

Environmental problems

Ecology must solve very complex multivariate problems with high variability and fuzziness. Therefore, pattern recognition techniques have been recently tested in various environmental data interpretations, for example, in:

- analysis of air pollution data on polycyclic aromatic hydrocarbons by a combination of interactive graphic technique and clustering technique (Gether and Seip, 1979),
- air pollution prediction (Pijpers, 1984),
- determination of aliphatic and aromatic hydrocarbons as estimated by gas chromatography (Saarinen, 1983),
- determination of polychlorinated biphenyls from "Aroclors" in environmental samples based on gas chromatography (Dunn et al., 1984),
- identification of oil-spill samples according to their fluorescence spectra (Killeen et al., 1981),
- determination of muscle components of blue mussels from polluted and non-polluted locations by gas chromatography (Kvalheim et al., 1983),
- assessment of oil contamination by analysis of paraffinic hydrocarbons of mussels (Kwan and Clark, 1981),
- differentiation of marine and fresh-water fish species according to the fatty acid composition (Forina et al., 1982b),
- monitoring water quality near an underground shale retort (Meglen and Erickson, 1983) and
- classification of human hair according to the trace metal composition (Pillay and Peisach, 1981).

Geochemical problems

Recently, several chemometric methods including pattern recognition methods have been applied for specific geological goals, especially for:

- interpretation of ground water data (Wegscheider and
  Leyden, 1979),
- evaluation of water flow from water quality data of each
  source (Gerlach et al., 1979),
- classification of mineral waters according to their
  elemental composition (Scarminio et al., 1982),
- assessment of the impact of the mine according to water
  quality data from groundwater flow from a strip coal
  mine (Brown et al., 1980),
- classification of coals (Kaufman et al., 1983),
- differentiation of Rocky Mountains coals by pyrolysis-
  mass spectroscopic data (Meuzelaar, 1982),
- classification of crude oils from similar lithostratigraphic situations on the basis of their elemental
  composition (Hitchon and Filby, 1984),
- classification in organic geochemistry (modelling of
  organic metamorphism) from chemical analysis of sedi-
  ments, oil shales, kerogens, fatty acids, oils and coals
  (Vuchev, 1983),
- classification of volcanic rocks according to their
  elemental composition (Bisani et al., 1983),
- study of geochemical dispersion around massive sulfide
  deposits (Donald, 1984),
- study of mineralization in granites, classification of
  diamonds, identification of sedimentary units and classification of coals from elemental composition (Watterson
  et al., 1983) and
- geochemical and geological applications using the SIMCA
  and the MACUP methods (Esbensen and Wold, 1983).

Archaeological problems

Recently, pattern recognition techniques have been used
in a series of different archaeological problems based

on chemical data, e.g., in:
- classification of potsherds according to concentrations
  of trace elements (Boulle and Peisach, 1979),
- location of historical glass findings according to their
  elemental composition (Nauer et al., 1980; Danzer et al.,
  1984),
- classification of bricks and wall slabs from a Roman
  settlement on the basis of elemental composition
  (Blasius et al., 1982) and
- classification of Roman coins produced in various mints
  of the Roman Empire based on their elemental composition
  (Borszéki et al., 1983).

Biochemical applications

Recently, interesting applications of chemical pattern
recognition in biochemistry have been described as shown
in the following examples of:
- determination of blood components by routine analytical
  methods in clinical laboratories (Jansen et al., 1981,
  1981a),
- differentiation of plasma from normal subjects and
  patients with acute lymphocytic leukemia by HPLC
  (Scoble et al., 1983),
- classification of liquid-chromatographic profiles of
  urine proteins in relation to a clinical assessment of
  the proteinuria caused by various renal disorders
  (Marshall et al., 1984),
- classification of normal and virus-infected serums by
  gas chromatography (Zlatkis et al., 1979),
- differentiation of urine from normal individuals and
  from individuals with diabetes mellitus by gas chromatography (McConnell et al., 1979),
- classification of urine samples from mass spectrometric

profiles (van der Greef et al., 1983),
- characterization of the functional state of the thyroid
  according to biochemical laboratory tests (Coomans et
  al.,1981b, 1981c, 1982c; Coomans and Massart, 1982,
  1982a, 1982b),
- classification of normal brain tissue from tumorous
  tissues by gas chromatography of brain biopsies (Jellum
  et al., 1981),
- classification of three types of human brain tissue by
  gas chromatography (Wold et al., 1981),
- differentiation of cholelithiosis and cholecystitis
  patients and normals by pyrolysis-mass spectrometry of
  human bile (Windig et al., 1983),
- detection of capsular polysaccharides in bacteria by
  pyrolysis-mass spectrometry (Windig et al., 1983),
- detection of cystic fibrosis heterozygotes from pyrolysis-gas chromatographic data (Pino et al., 1985),
- elucidation of metabolic pathways for a biosynthesis of
  fatty acids in the milk of lactating goats according to
  the content of these acids (Massart-Leën and Massart,
  1981),
- in clinical chemistry, where the determination of a
  large number of components in a small number of samples
  is a common case, for the analysis of curves called
  "attention function" (Goldschmidt et al., 1983) and
- for the analysis of a chemical communication (Smith et
  al., 1985).

CHAPTER 4
Recent Trends in Chemical Pattern Recognition

In the past years chemical pattern recognition has been one of the most quickly developing parts of chemometrics (see Introduction), as is clear from the results described in more than 350 publications published from 1979 to 1985 (see References). These results form a reasonably representative (not exhaustive) sample, for a discussion about the main recent trends in chemical pattern recognition, as well as about possible developments in the near future.

4.1 RECENT TRENDS IN CHEMICAL DATA PROCESSING

A remarkable interest in processing large chemical data sets can be observed (Section 1.1). One can expect further intensive progress in this field mainly from two points of view:

(a) the requirement of a more economical exploitation of information contained in the data and

(b) the necessity of relieving chemists from an increasing information burden in favour of their creative activity.

In the author's opinion, the progress in processing chemical data by the use of pattern recognition approaches. will be characterized by

(1) even more sophisticated procedures for processing
     very large data bases gathered during systematic
     (and often very expensive) studies made by institutes
     or organizations so that information may be access-
     ible in a highly concise form and
(2) development of pattern recognition algorithms work-
     ing in real time and suitable for on-line combination
     with modern analytical instruments.

The main advantage of computerized data processing is its
velocity and reliability in comparison with those of man
-made processing. Therefore, the use of computers in pattern recognition processing of chemical data seems to be
the main trend also in the near future.

4.2 RECENT TRENDS IN THE MODELLING OF CHEMICAL SYSTEMS

The use of pattern recognition for physical modelling of
chemical systems with the goal of prediction is a more
problematic task (Section 1.2). It should be underlined
that any generalization could be valid only for a closed
set of objects represented by features of a fundamental
physico-chemical character. In such cases the pattern
recognition analysis can critically assess the existing
hypotheses in an unbiased way. Moreover, the use of features of a theoretical character could provide some new
impulses for the innovation of theories; this exciting
possibility is in its infancy and its development in the
near future seems to be a very interesting topic for
discussion. Irrespective of the sophistication level of
algorithms, the success of chemical modelling by pattern
recognition is, and always will be, dependent on critical
and careful use of chemist's a priori knowledge. Therefore, a man-computer dialogue is an effective way forward
in this field.

4.3 RECENT TRENDS IN CHEMICAL PATTERN RECOGNITION
METHODOLOGY

The recent methodological developments (Chapter 2) orig-
inate either as adaptations of methods developed orig-
inally for other disciplines or as new methods devised
in connection with chemical pattern recognition.

A great variety of supervised pattern-recognition classification methods (Section 2.1) have been used for very
different chemical problems. This is a natural consequence
of the non-generality of pattern recognition methods,
when a method very efficient for a given chemical system
is not the optimum one for another chemical system. For
chemical modelling, the pattern recognition methods based
on mathematical modelling of classes seem to be suitable.
The importance of cluster analysis and other unsupervised methods in chemistry (Section 2.2) has been recently emphasized and some new versatile methods have been
developed and tested.

Lately, a significant interest has been also focused on
feature selection methods (Section 2.3), evidently thanks
to their importance for both the reduction of dimensionality and the improvement of classification.

In this connection, also visualization methods (Section
2.4) have been intensively studied because they utilize
to full effect the extraordinary recognition ability of
the human being. Moreover, other human senses than eyes
can be very effective in a recognition process, as has
been shown in the case of audio-display. The use of some
other display principles could be also fruitful in chemistry; thus, a method that creates a human face from chemical features could give a representation of chemical objects in a form of some type of human face. The role of
visualization in chemist-computer interaction has not yet
been sufficiently exploited.

4.4 RECENT TRENDS IN APPLICATIONS OF CHEMICAL PATTERN RECOGNITION

A variety of applications of chemical pattern recognition (Chapter 3) have been described in recent literature (see References). Chemical data (Section 3.1) are obtained either during an every-day laboratory job or they are extracted from literature. The former case is very frequent in routine analytical and biochemical laboratories. In the latter case, a very well organized chemical literature can serve as a valuable source of chemical data, which can be then analysed by pattern recognition to yield information contained in them and hitherto unexploited. A highly convenient advantage of chemical pattern recognition is the possibility to analyse, after an appropriate preprocessing, data of different types measured on the same object; such a set of heterogeneous (multi-source) data can yield, in many cases, correct recognition of the pattern, which cannot be achieved using a homogeneous (single-source) data set only.

The goal of a pattern recognition analysis using chemical data is mainly the classification of patterns into physical, chemical and biological classes (Section 3.2). A physical classification property can be successfully modelled if the features representing the chemical pattern are related to the physical behaviour. Examples have been recently very rare, in spite of the potential usefulness for objective assessment of physical models. The "pure" chemical pattern recognition, in which one seeks for a chemical classification property from chemical features, represents a more frequent application. Most frequent have been chemical structure-biological activity studies, which, naturally, are the most attractive, but also the most dangerous ones. Triviality and/or statistical invalidity can often be found in these applications. Nevertheless, the author's opinion is that this controversial and important problem will be a major application of pattern recognition in the near future.

In conclusion it should be said, that the maturing of chemical pattern recognition as a "direct mathematization of chemistry" (Malissa, 1984) in the near future can be expected as "perhaps the most utilitarian of the chemometric methods" (Howery and Hirsch, 1983). The invasion of microcomputers into chemical laboratories will, naturally, cause a broader use of suitable pattern recognition methods in daily practice as an efficient tool for evaluation and control of experiments.

Procedures developed in connection with other scientific disciplines will be tested in chemistry. For example, syntactic (structural) pattern recognition methods (Fu, 1982) could be useful for chemistry as a highly formalized discipline dealing with structures. Moreover, systems called "expert systems", using a priori knowledge of an expert for solution of very complex problem, have not been so far commonly used in chemistry (see Musch et al., 1985 and references therein).

Henceforth, pattern recognition should be considered as an open rapidly developing system which promises, in a short time, a broad use of more efficient methods. As a typical open system, chemical pattern recognition requires, for its successful development, active contacts between people interested in the field, which can serve for encouragement and criticism. This is also the aim of this monograph.

References

Aartsma, T.J., Gouterman, M., Jochum, C., Kwiram, A.L., Pepich, B.V., Williams, L.D. (1982). Porphyrins. 43. Triplet sublevel emission of platinum tetrabenzo-porphyrin by spectrothermal principal component decomposition. J.Amer.Chem.Soc. 104, 6278-6283.

Abe, H., Yamasaki, T., Fujiwara, I., Sasaki, S. (1981). Computer-aided structure elucidation methods. Anal. Chim.Acta 133, 499-506.

Åberg, E.R., Gustavsson, A.G.T. (1982). Design and evaluation of Modified Simplex methods. Anal.Chim.Acta 144, 39-53.

Adamson, G.W., Bawden, D. (1975). A method of structure-activity correlation using Wiswesser line notation. J.Chem.Inf.Comput.Sci. 15, 214-220.

Adamson, G.W., Bawden, D. (1976). An empirical method of structure-activity correlation for polysubstituted cyclic compounds using Wiswesser line notation. J.Chem. Inf.Comput.Sci. 16, 161-165.

Adamson, G.W., Bawden, D. (1980). Automated additive modelling techniques applied to thermochemical property estimation. J.Chem.Inf.Comput.Sci. 20, 242-246.

Adamson, G.W., Bawden, D. (1981). Comparison of hierarchical cluster analysis techniques for automatic classification of chemical structures. J.Chem.Inf.Comput.Sci. 21, 204-209.

150

Ahrens, H., Läuter, J. (1974). Mehrdimenzionale Varianz-analyse. Akademie-Verlag, Berlin.

Aishima, T., Nagasawa, M., Fukushima, D. (1979). Differentiation of the aroma quality of soy sauce by statistical evaluation of gas chromatographic profiles. J.Food Sci. 44, 1723-1731.

Albano, C., Dunn, W.,III, Edlund, U., Johansson, E., Nordén, B., Sjöström, M., Wold, S. (1978). Four levels of pattern recognition. Anal.Chim.Acta 103, 429-443.

Alunni, S., Clementi, S., Edlund, U., Johnels, D., Hellberg, S., Sjöström, M., Wold, S. (1983). Multivariate data analysis of substituent descriptors. Acta Chem. Scand. B37, 47-53.

Amirova, Z.K., Masagutov, R.M., Morozov, B.F., Spivak, S.I. (1980). Analysis of selection of catalysts by means of pattern recognition method. Kinet.Katal. 21, 1174-1177.

Amirova, Z.K., Masagutov, R.M. (1982). Formal evaluation of the classification of heterogeneous catalytic reactions with the participation of molecular oxygen. Kinet.Katal. 23, 666-669.

Anderegg, R.J. (1981). Selective reduction of mass spectral data by isotope cluster chromatography. Anal.Chem. 53, 2169-2171.

Anderson, T.W., Bahadur, R.R. (1962). Classification into two multivariate normal distributions with different covariance matrices. Ann.Math.Stat. 33, 420-431.

Andrews, H.C. (1972). Introduction to Mathematical Techniques in Pattern Recognition. Wiley, New York.

Arunachalam, J., Gangadharan, S., Yegnasubramanian, S. (1983). Trace Anal.Technol.Develop., Pap.Int.Symp., 1st 1981, pp.174-181. Wiley, New York.

Arunachalam, J., Gangadharan, S. (1984). Feature extraction from spectral and other data by the principal components and discriminant function techniques. Anal.Chim.Acta 157, 245-260.

Austel, V. (1983). Drug design. Design of test series by combined application of 2^n factorial schemes and pattern recognition techniques. Pharmacochem.Libr. <u>6</u>, 223-229.

Bailey, A. (1979), in The Medical and Biological Application of Mass Spectrometry. (Edited by J.P. Payne, J.A. Bushman and D.W. Hill). p.93, Academic Press, London.

Barenboim, G.M., Brikenstein, V.Kh., Ovchinnikov, A.A., Pitina, L.R., Terekhina, A.I., Shamovskii, I.L. (1982). Use of luminiscence-absorption analysis and pattern recognition theory for evaluating the biological activity of chemical compounds. 1. Quantum biochemical interpretation of absorption spectra of steroidal hormones. Khim.-Farm.Zh. <u>16</u>, 823-826.

Bawden, D. (1983), in Quantitative Approaches to Drug Design. (Edited by J.C. Dearden). p.64, Elsevier, Amsterdam.

Bawden, D. (1984). DISCLOSE: an integrated set of multivariate display procedures for chemical and pharmaceutical data. Anal.Chim.Acta <u>158</u>, 363-368.

Bawden, D,m Brock, A.M. (1985). Chemical toxicology searching. A comparative study of online data-bases. J.Chem. Inf.Comput.Sci. <u>25</u>, 31-35.

Bawden, D., Fisher, J.D. (1985a). A note on measures of screening effectiveness in chemical substructure searching. J.Chem.Inf.Comput.Sci. <u>25</u>, 36-38.

Belozerskikh, V.A., Dobrotvorskii, A.M. (1980). Use of mathematical methods of the pattern recognition theory for the prediction of alloy properties. Deposited Doc. SPSTL 870 khp-D80, 122-129. Chem. Abstr. (1982), <u>97</u>, 149 276.

Bertsch, W., Mayfield, H., Thomason, M.M. (1981). Application of pattern recognition to high resolution GC and GC/MS. 1. Basic studies. Proc. Int. Symp. Capillary Chromatogr., 4th (Edited by R.E. Kaiser). p.313. Chem. Abstr. (1982), <u>97</u>, 229 483.

Betteridge, D., Lilley, T., Cudby, M.E.A. (1980). Acoustic emissions from polymers. 2. Use of pattern recognition methods. Anal.Proc. 17, 434-436.

Bink, J.C.W.G., van't Klooster, H.A. (1983). Classification of organic compounds by infrared spectroscopy with pattern recognition and information theory. Anal.Chim. Acta 150, 53-59.

Bisani, M.L., Clementi, S., Wold, S. (1982). Chemical measurement elements. 1. Multivariate statistical analysis in chemistry. Chim.Ind. (Milan) 64, 655-665.

Bisani, M.L., Clementi, S., Wold, S. (1982a). The elements of chemometrics. 2. The SIMCA method. Chim.Ind. (Milan) 64, 727-741.

Bisani, M.L., Faraone, D., Clementi, S., Esbensen, K.H., Wold, S. (1983). Principal components and partial least squares analysis of the geochemistry of volcanic rocks from the Aeolian archipelago. Anal.Chim.Acta 150, 129-143.

Blasius, E., Wagner, H., Braun, H., Krumbholtz, R., Thimmel, B. (1982). Archaeometrical studies of Roman bricks and wall slabs. 1. Graphic and computational evaluation of analytical data. Z.Anal.Chem. 310, 98-107.

Blomquist, G., Johansson, E., Söderström, B., Wold, S. (1979). Data analysis of pyrolysis chromatograms by means of SIMCA pattern recognition. J.Anal.Appl. Pyrolysis, 1, 53-65.

Blomquist, G., Johansson, E., Söderström, B., Wold, S. (1979a). Classification of fungi by pyrolysis-gas chromatography-pattern recognition. J.Chromatogr. 173, 19-32.

Blomquist, G., Johansson, E., Söderström, B., Wold, S. (1979b). Reproducibility of pyrolysis-gas chromatographic analysis of the mould Penicillium brevicompactum. J.Chromatogr. 173, 7-17.

Bodor, N., Harget, A.J., Phillips, E. (1983). Structure-activity relationships in the antiinflammatory steroids: a pattern recognition approach. Croatica Chem.Acta 56, 175-183.

Bodor, N., Harget, A.J., Phillips, E.W. (1983a). Structure -activity relationships in the antiinflammatory steroids: a pattern recognition approach. J.Med.Chem. 26, 318-328.

Bonchev, D. (1983). Information Theoretic Indices for Characterization of Chemical Structures. Research Studies Press, Chichester.

Boon, J.J., Tom, A., Brandt, B., Eijkel, G.B., Kistemaker, P.G. (1984). Mass spectrometric and factor discriminant analysis of complex organic matter from the bacterial culture environment of Bacteroides gingivalis. Anal. Chim.Acta 163, 193-205.

Borszéki, J., Inczédy, J., Gegus, E., Óvári, F. (1983). Evaluation by pattern recognition methods of laser micro spectral analysis data of Roman coins. Fresenius Z.Anal.Chem. 314, 410-413.

Bos, M. (1979). The learning machine in quantitative chemical analysis. 2. Potentiometric titrations of mixtures of three bases. Anal.Chim.Acta 112, 65-73.

Boulle, G.J., Peisach, M. (1979). Trace element analysis of archaeological materials and the use of pattern recognition methods to establish identity. J.Radioanal. Chem. 50, 205-215.

Božinovski, S. (1985). A representation theorem for linear pattern classifier training. IEEE Trans.Systems, Man, Cybern. 15, 159-161.

Brill, J.H., Mayfield, H.T., Mar, T., Bertsch, W. (1985). Use of computerized pattern recognition in the study of the cuticular hydrocarbons of imported fire ants. 1. Introduction and characterization of the cuticular hydrocarbon patterns of Solenopsis invicta and S. Richteri. J.Chromatogr. 349, 31-38.

Brill, J.H., Mar, T., Mayfield, H.T., Bertsch, W. (1985a).

Use of computerized pattern recognition in the study of the cuticular hydrocarbons of imported fire ants. 2. Comparison of the cuticular hydrocarbon patterns between different colonies of Solenopsis Richteri. J.Chromatogr. 349, 39-48.

Brissey, G.F., Spencer, R.B., Wilkins, C.L. (1979). High-speed algorithm for simplex optimization calculations. Anal.Chem. 51, 2295-2297.

Brown, S.D., Skogerboe, R.K., Kowalski, B.R. (1980). Pattern recognition assessment of water quality data: coal strip mine drainage. Chemosphere 9, 265-276. Chem. Abstr. (1980), 93, 209 953.

Brugger, W.E., Stuper, A.J., Jurs, P.C. (1976). Generation of descriptors from molecular structures. J.Chem.Inf. Comput.Sci. 16, 105-110.

Brunner, T.R., Wilkins, C.L., Lam, T.F., Soltzberg, L.J., Kaberline, S.L. (1976). Simplex pattern recognition applied to carbon-13 nuclear magnetic resonance spectrometry. Anal.Chem. 48, 1146-1150.

Byers, W.A., Perone, S.P. (1979). Computerized pattern recognition applied to nickel-cadmium cell lifetime prediction. J.Electrochem.Soc. 126, 720-725.

Byers, W.A., Perone, S.P. (1980). K-Nearest Neighbour rule in weighting measurements for pattern recognition. Anal.Chem. 52, 2173-2177.

Byers, W.A. (1981). Qualitative organic electroanalysis using computerized pattern recognition. Diss.Abstr. Int.B 42, 4403.

Byers, W.A., Perone, S.P. (1983). Generation of an electrochemical data base for pattern recognition. Anal.Chem. 55, 615-620.

Byers, W.A., Freiser, B.S., Perone, S.P. (1983a). Structural and activity characterization of organic compounds by electroanalysis and pattern recognition. Anal.Chem. 55, 620-625.

Carbone, D., Musumarra, G., Occhipinti, S., Scarlata, G., Wold, S. (1983). Thin-layer chromatographic behaviour of amines and related compounds: multivariate analysis of R_f data in chloroform-ethyl acetate mixture. Ann. Chim. (Italy), 73, 183-192.

Carpenter, R.C., Till, C. (1984). Analysis of small samples of brasses by inductively coupled plasma-optical emission spectrometry and their classification by two pattern-recognition techniques. Analyst 109, 881-884.

Chapman, J.R. (1978). Computers in Mass Spectrometry. p.150, Academic Press, London.

Chapman, N.B., Shorter, J. (Editors). (1972). Advances in Linear Free Energy Relationships. Plenum, London.

Chapman, N.B., Shorter, J. (Editors) (1978). Correlation Analysis in Chemistry. Plenum, London.

Chemleva, T.A., Elfimova, T.L., Shubina, M.D., Isaeva, E.V., Evseev, A.M. (1982). The possibility of the classification of azomethines in relation to their photochromic properties by the pattern recognition technique. Vestn.Mosk.Univ., Khim. 23, 341-345.

Chen, B.-K., Horvath, C., Bertino, J.R. (1979). Multi-variate analysis and QSAR. Inhibition of dihydrofolate reductase and thymidylate synthetase by quinazolines. J.Med.Chem. 22, 483-491.

Chen, N.Y., Xie, L.M., Shi, T.S., Jiang, N.X., Li, Q.Z. (1981). Computerized pattern recognition applied to chemical bond research. Model and methods of computation. Scientia Sinica 24, 1528-1535.

Chien, M. (1985). Analysis of complex mixtures by gas chromatography using a pattern recognition. Anal.Chem. 57, 348-352.

Chou, J.T., Jurs, P.C. (1979). Computer assisted structure-activity studies of chemical carcinogens. An N-nitroso compound data set. J.Med.Chem. 22, 792-797.

Chou, J.T., Jurs, P.C. (1979a). Computer-assisted computation of partition coefficients from molecular structures using fragment constants. J.Chem.Inf.Comput.Sci. 19, 172-178.

Clark, H.A., Jurs, P.C. (1979). Classification of crude oil gas chromatograms by pattern recognition techniques. Anal.Chem. 51, 616-623.

Cleij, P., Dijkstra, A. (1979). Information theory applied to qualitative analysis. Fresenius Z.Anal.Chem. 298, 97-109.

Clerc, J.T., Székely, G. (1984). Pattern recognition for structure elucidation. Hypotheses as to why today's approaches cannot work, in Modern Trends in Analytical Chemistry, Part B.(Edited by E. Pungor, G.E. Veress and I. Buzás). p.49, Akadémiai Kiadó, Budapest.

Coomans, D., Jonckheer, M., Massart, D.L., Broeckaert, I., Blockx, P. (1978). The application of linear discriminant analysis in the diagnosis of thyroid diseases. Anal.Chim.Acta 103, 409-415.

Coomans, D., Massart, D.L., Kaufman, L. (1979). Optimization by statistical linear discriminant analysis in analytical chemistry. Anal.Chim.Acta 112, 97-122.

Coomans, D., Massart, D.L., Broeckaert, I., Tassin, A. (1981). Potential methods in pattern recognition. 1. Classification aspects of the supervised method ALLOC. Anal.Chim.Acta 133, 215-225.

Coomans, D., Massart, D.L. (1981a). Potential methods in pattern recognition. 2. CLUPOT - an unsupervised pattern recognition technique. Anal.Chim.Acta 133, 225-239.

Coomans, D., Derde, M., Massart, D.L., Broeckaert, I. (1981b). Potential methods in pattern recognition. 3. Feature selection with ALLOC. Anal.Chim.Acta 133, 241-250.

Coomans, D., Massart, D.L., Broeckaert, I. (1981c). Potential methods in pattern recognition. 4. A combination of ALLOC and statistical linear discriminant analysis. Anal.Chim.Acta 132, 69-74.

Coomans, D., Massart, D.L. (1982). Alternative k-nearest neighbour rules in supervised pattern recognition. 1. k-Nearest neighbour classification by using alternative voting rules. Anal.Chim.Acta 136, 15-27.

Coomans, D., Massart, D.L. (1982a). Alternative k-nearest neighbour rules in supervised pattern recognition. 2. Probabilistic classification on the basis of the kNN method modified for direct density estimation. Anal. Chim.Acta 138, 153-165.

Coomans, D., Massart, D.L. (1982b). Alternative k-nearest neighbour rules in supervised pattern recognition. 3. Condensed nearest neighbour rules. Anal.Chim.Acta 138, 167-176.

Coomans, D., Massart, D.L., Broeckaert, I. (1982c). Potential methods in pattern recognition. 5. ALLOC, action-oriented decision making. Anal.Chim.Acta 134, 139-151.

Crawford, L.R., Morrison, J.D. (1968). Computer methods in analytical mass spectrometry. Empirical identification of molecular class. Anal.Chem. 40, 1469-1474.

Danzer, K., Singer, R., Mäurer, F., Flórián, K., Zimmer, K. (1984). Multidimensional variance and discriminance analysis of spectrographical data of glass bead findings. Fresenius Z.Anal.Chem. 318, 517-521.

Danzer, K., Singer, R. (1985). Application of pattern recognition methods for investigation of chemical homogeneity of solids. Microchim.Acta 1, 219-226.

DeBeer, J.O., Heyndvickx, A.M. (1982). Numerical taxonomy of common phases for gas-liquid chromatography using chloroperoxy alkyl esters as test substances. J.Chromatogr. 235, 337-349.

DeJong, E., Pijpers, F.W. (1985). Investigations of the secondary structure of 5S-ribonucleic acids by means of pattern recognition. Anal.Chim.Acta 167, 97-109.

Delaney, M.F., Denzer, P.C., Barnes, R.M., Uden, P.C. (1979). Pattern recognition approach to vapour phase infrared spectra interpretation for gas chromatography. Anal.Lett. 12, 963-978.

Delaney, M.F. (1984). Chemometrics. Anal.Chem. 56, 261R-277R.

Delaney, M.F., Hallowell, J.R.,Jr., Warren, F.V.,Jr. (1985). Optimization of a similarity metric for library searching of highly compressed vapour-phase infrared spectra. J.Chem.Inf.Comput.Sci. 25, 27-30.

DePalma, R.A., Perone, S.P. (1979). On-line pattern recognition of voltammetric data. Peak multiplicity classification. Anal.Chem. 51, 825-828.

DePalma, R.A., Perone, S.P. (1979a). Characterization of heterogeneous kinetic parameters from voltammetric data by computerized pattern recognition. Anal.Chem. 51, 829-832.

Derde, M.P., Massart, D.L. (1982). Extraction of information from large data sets by pattern recognition. Fresenius Z.Anal.Chem. 313, 484-495.

Derde, M.P., Coomans, D., Massart, D.L. (1982a). Effect of scaling on class modelling with the SIMCA method. Anal.Chim.Acta 141, 187-192.

Derde, M.P., Massart, D.L., Ooghe, W., DeWaele, A. (1983). Use of pattern recognition display techniques to visualize data contained in complex data-bases. J.Automat. Chem. 5, 136-145.

Diehr, G. (1985). Evaluation of a branch and bound algorithm for clustering. SIAM J.Sci.Stat.Comput. 6, 268-284.

Domokos, L., Pretsch, E., Mändli, H., Könitzer, H., Clerc, J.T. (1980). Cluster analysis of mass-spectrometric data. Fresenius Z.Anal.Chem. 304, 241-249.

Domokos, L., Frank, I. (1981). Orthogonal transformation

for feature extraction in chemical pattern recognition.
Anal.Chim.Acta 133, 261-270.

Domokos, L., Frank, I., Matolcsy, G., Jalsovszky, G.
(1983). Pattern recognition applied to vapour-phase
infrared spectra. Anal.Chim.Acta 154, 181-189.

Domokos, L., Henneberg, D., Weimann, B. (1984). Computer
-aided identification of compounds by comparison of
mass spectra. Anal.Chim.Acta 165, 61-74.

Domokos, L., Henneberg, D. (1984a). A correlation method
in library search. Anal.Chim.Acta 165, 75-86.

Donald, A.S. (1984). Application of discriminant analysis
to a study of geochemical dispersion around massive
sulfide deposits in Superior Province. Diss. Abstr.
Int.B 44, 3689.

Dove, S., Franke, R., Mndshojan, O.L., Schkuljev, W.A.,
Chashakjan, L.W. (1979). Discriminant-analytical investigation on the structural dependence of hyper and hypo-
glycemic activities in a series of substituted o-tolyl-
sulfonyl-(thio)ureas. J.Med.Chem. 22, 90-95.

Dove, S., Streich, W.J., Franke, R. (1980). On the rational selection of test series. 2. Two-dimensional
mapping of intraclass correlation matrices. J.Med.Chem.
23, 1456-1459.

Drack, H., Kosina, S., Grasserbauer, M. (1979). Compound
-specific in-situ microanalysis with an electron micro-
probe-characterization of X-ray valence band spectra
with band vectors and representative space transformation. Fresenius Z.Anal.Chem. 295, 30-35.

Duewer, D.L., Koskinen, J.R., Kowalski, B.R. (1975).
Documentation for ARTHUR Version 1-8-75. Chemometric
Society Report No.2, Laboratory for Chemometrics.

Dunn, W.J.,III, Wold, S. (1980). Relationship between
chemical structure and biological activity modelled by
SIMCA pattern recognition. Bioorgan.Chem. 9, 505-523.

Dunn, W.J.,III, Wold, S. (1980a). Structure-activity

analysed by pattern recognition. The asymmetric case.
J.Med.Chem. 23, 595-599.

Dunn, W.J.,III, Wold, S. (1981). The cancinogenicity of
N-nitroso compounds: a SIMCA pattern recognition study.
Bioorgan.Chem. 10, 29-45.

Dunn, W.J.,III, Wold, S. (1981a). An assessment of
carcinogenicity of N-nitroso compounds by the SIMCA
method of pattern recognition. J.Chem.Inf.Comput.Sci.
21, 8-13.

Dunn, W.J.,III, Wold, S., Edlund, U., Hellberg, S.,
Gasteiger, J. (1982). Structure-activity relationships
between data from a battery of biological tests and an
ensemble of biological descriptors: the PLS method.
Techn.Report No.2. Research Group for Chemometrics,
Umeå University.

Dunn, W.J.,III, Lins, C., Kumar, G., Maninaran, T.,
Grigoras, S., Edlund, U., Wold, S. (1982a). Substituent
effects on B-13 and N-15 chemical shifts in triazenes
studies by principal components multivariate data analysis. Org.Magn.Reson. 21, 450-456.

Dunn, W.J.,III, Wold, S. (1983). The use of SIMCA pattern
recognition in predicting the carcinogenicity of potential environmental pollutants. Struct.-Act. Correl.
Predict. Tool Toxicol. (Edited by L. Goldberg). p.141,
Hemisphere, Washington.

Dunn, W.J.,III, Stalling, D.L., Schwartz, T.R., Hogan, J.
W., Petty, J.D., Johansson, E., Wold, S (1984). Pattern
recognition for classification and determination of
polychlorinated biphenyls in environmental samples.
Anal.Chem. 56, 1308-1313.

Eckschlager, K., Štěpánek, V. (1979). Information Theory
as Applied to Chemical Analysis. p.155, Wiley, New York.

Eckschlager, K., Horsák, I., Kodejš, Z., Ksandr, Z.,
Matherny, M., Obrusník, I., Wičar, S. (1983). Application of computers in analytical chemistry, in Compre-

hensive Analytical Chemistry, Vol.18 (edited by G. Svehla). p.357, Elsevier, Amsterdam.

Eckschlager, K., Štěpánek, V. (1985). Analytical Measurement and Information. Advances in the Information Theoretic Approach to Chemical Analysis. Research Studies Press, Chichester.

Edlund, U., Wold, S. (1980). Interpretation of NMR substituent parameters by the use of a pattern recognition approach. J.Magn.Reson. 37, 183-194.

Edward, J.T., Sjöström, M., Wold, S. (1981). Multivariate analysis of data for the ionization of weak acids in water-dimethyl sulfoxide solvent mixtures. Can.J.Chem. 59, 2350-2357.

Engman, H., Mayfield, H.T., Mar, T., Bertsch, W. (1984). Classification of bacteria by pyrolysis-capillary column gas chromatography-mass spectrometry and pattern recognition. J.Anal.Appl.Pyrolysis 6, 137-156.

Esbensen, K.H., Wold, S. (1983). SIMCA, MACUP, SELPLS, GDAM, SPACE and UNFOLD: the way towards regionalized principal analysis and subconstrained N-way decomposition - with geological illustration. Proc. Conf. Appl. Statistics, Stavanger. (Edited by O.J. Christie). p.11.

Esbensen, K.H., Kaufman, L., Massart, D.L. (1984). Interobject structure of unchanged iron meteorites as revealed by advanced clustering: methods and chemical features. Meteoritics 19, 95-109.

Everitt, B. (1974). Cluster Analysis. Heinemann, London.

Fainzil'berg, L.S. (1980). New approach to thermal analysis of molten steel for carbon. Izv.Vyssh.Uchebn.Zaved., Chern.Metall., p.113. Chem. Abstr. (1980), 93, 87 788.

Farkas, M., Markos, J., Szepesváry, P., Bartha, I., Szalontai, G., Simon, Z. (1980). Computerized analytical system (ASSIGNER) for structure determination of organic compounds. Magy.Kem.Lapja 35, 605-612. Chem. Abstr. (1981), 95, 60 708.

Farkas, M., Markos, J., Szepesváry, P., Bartha, I.,
Szalontai, G., Simon, Z. (1981). Computerized analytical
system (ASSIGNER) for structure determination of organic
compounds. Anal.Chim.Acta 133, 19-29.

Fellous, R., Lafaye de Micheaux, D., Lizzani-Cuvelier, L.,
Luft, R. (1981). Behaviour of stationary phases. 1.
Linear relationships between retention data of benzene
derivatives. J.Chromatogr. 213, 223-242.

Fellous, R., Lafaye de Micheaux, D., Lizzani-Cuvelier, L.,
Luft, R. (1982). Chromatographic behaviour of stationary
phases. 2. Application of factor analysis. J.Chromatogr.
248, 35-47.

Fellous, R., Lizzani-Cuvelier, L., Luft, R., Lafaye de
Micheaux, D. (1983). Data analysis in gas-liquid
chromatography of benzene derivatives. Anal.Chim.Acta
154, 191-201.

Fisher, R.A. (1936). The use of multiple measurements in
taxonomic problems. Ann.Eugen. 7, 179.

Foley, D.H., Sammon, J.W.,Jr. (1975). An optimal set of
discriminant vectors. IEEE Trans.Comput. 24, 281-289.

Forina, M., Armanino, C. (1982). Eigenvector projection
and simplified nonlinear mapping of fatty acid content
of Italian olive oils. Ann.Chim. (Rome), 72, 127-141.

Forina, M., Tiscornia, E. (1982a). Pattern recognition
methods in the prediction of Italian olive oil origin
by their fatty acid content. Ann.Chim. (Rome) 72,
143-155.

Forina, M., Armanino, C., Lanteri, S. (1982b). Fatty acid
content of aquatic animals: a chemometric study. Riv.
Soc.Ital.Sci.Aliment. 11, 15-22.

Frank, I.E., Kowalski, B.R. (1982). Chemometrics. Anal.
Chem. 54, 232R-243R.

Frankel, D.S. (1984). Pattern recognition of Fourier
transform infrared spectra of organic compounds. Anal.
Chem. 56, 1011-1014.

Fu, K.S. (1982). Syntactic Pattern Recognition and Applications. Prentice-Hall, Englewood Cliffs.

Fusek, J., Štrouf, O. (1979). Estimation of the stability of complex hydrides by a pattern-recognition classification method. Coll.Czechoslov.Chem.Commun. 44, 1362-1369.

Fusek, J., Štrouf, O. (1980). Modelling of complex hydrides stability by a pattern recognition method. Commun. Czech.-Pol Colloq. Chem. Thermodyn. Phys. Org. Chem., 2nd, p.160.

Fusek, J., Štrouf, O., Kuchynka, K. (1983). Approximation of adsorption heats of carbon monoxide on transition metals by means of an empirical model. Coll.Czechoslov. Chem.Commun. 48, 2735-2739.

Gaarenstroom, S.W. (1979). Chemical characterization from carbon Auger spectra by application of pattern recognition and factor analysis. J.Vac.Sci.Technol. 16, 600-604.

Garozzo, D., Montaudo, G. (1985). Identification of polymers by library search of pyrolysis mass spectra and pattern recognition analysis. J.Anal.Appl.Pyrolysis 9, 1-17.

Gerlach, R.W., Kowalski, B.R., Wold, H.O.A. (1979). Partial least-squares path modelling with latent variables. Anal.Chim.Acta 112, 417-421.

Gether, J., Seip, H.M. (1979). Analysis of air pollution data by the combined use of interactive graphic presentation and a clustering technique. Atmos.Envir. 13, 87-96.

Goldschmidt, H.M.J., Leijten, J.F., Scholten, M.N.M. (1983). Modelling component combination by means of attention function scores. Anal.Chim.Acta 150, 207-217.

Golender, V.E., Rozenblit, A.B. (1980), in Drug Design, Vol.9. (Edited by E.J. Ariens). p.299, Academic Press, New York.

Gribov, L.A. (1980). Application of artificial intelligence systems in molecular spectroscopy. Anal.Chim.
Acta 122, 249-256.

Gribov, L.A. (1984). The problem of spectroscopic study
of polyatomic molecules based on computers, in Modern
Trends in Analytical Chemistry, Part B. (Edited by
E. Pungor, G.E. Veress and I. Buzás). p.59, Akadémiai
Kiadó, Budapest.

Gustavsson, A., Sundkvist, J.E. (1985). Design and optimization of modified simplex methods. Anal.Chim.Acta
167, 1-10.

Habbema, J.D.F. (1983). Some useful extensions of the
standard model for probabilistic supervised pattern
recognition. Anal.Chim.Acta 150, 1-10.

Hangac, G., Wieboldst, R., Lam, R., Isenhour, T.L. (1982).
Compression of an infrared spectral library by Karhunen
-Loève transformation. Appl.Spectrosc. 36, 40-47.

Hanna, D.A., Hangac, C., Hohne, B.A., Small, G.W.,
Wieboldt, R.C., Isenhour, T.L. (1979). A comparison
of methods used for the reconstruction of GC/FT-IR
chromatograms. J.Chromatogr.Sci. 17, 423-427.

Hanna, A., Marshall, J.C., Isenhour, T.L. (1979a). A GC/
FT-IR compound identification system. J.Chromatogr.Sci.
17, 434-440.

Harper, A.M., Duewer, D.L., Kowalski, B.R., Fasching, J.L.
(1977). ARTHUR and experimental data analysis. Heuristic
use of a polyalgorithm, in Chemometrics: Theory and
Application. (Edited by B.R. Kowalski). ACS Symp. Series,
Vol.52, p.14, Amer.Chem.Soc., Washington.

Harper, A.M. (1985), in Pyrolysis in Gas Chromatography
in Polymer Analysis. (Edited by S.A. Liebman and E.J.
Levy). p.373, Marcel Dekker, New York.

Havel, J., Meloun, M. (1985). Multiparametric curve fit-
ting. 7. Determination of the number of complex species
by factor analysis of potentiometric data. Talanta 32,

171-175.

Heller, S., Potenzone, R. (Editors). (1983). Computer
Applications in Chemistry. Elsevier, New York.

Henry, D.R., Block, J.H. (1979). Classification of drugs
by discriminant analysis using fragment molecular
connectivity values. J.Med.Chem. 22, 465-472.

Henry, D.R., Block, J.H. (1980). Pattern recognition of
steroids using molecular connectivity. J.Pharm.Sci.
69, 1030-1034.

Henry, D.R., Jurs, P.C., Denny, W.A. (1982). Structure-
antitumour activity relationships of 9-anilinoacridines
using pattern recognition. J.Med.Chem. 25, 899-908.

Hermans, J., Habbema, J.D.F. (1976). Manual for the ALLOC
-discriminant analysis program. Department of Medical
Statistics, University of Leiden.

Hitchon, B., Filby, R.H. (1984). Use of trace elements for
classification of crude oils into families. Example from
Alberta, Canada. AAPG Bulletin 68, 838-849.

Hodes, L. (1979), in Computer-Assisted Drug Design.
(Edited by E.C. Olson and R.E. Christoffersen). ACS
Symp. Series No.112, p.583, Am.Chem.Soc., Washington.

Hodes, L. (1981). Computer-aided selection of compounds
for antitumour screening. J.Chem.Inf.Comput.Sci. 21,
128-132.

Hodes, L. (1981a). Selection of molecular fragment features for structure-activity studies in antitumour
screening. J.Chem.Inf.Comput.Sci. 21, 132-136.

Hohne, B.A., Hangac, G., Small, G.W., Isenhour, T.L.
(1981). An on-line class-specific GC/FT-IR reconstruction from interferometric data. J.Chromatogr.Sci. 19,
283-289.

Honeybourne, C.L., Smith, G.E. (1982). The digitization
of rotational fine structure as a data base for pattern
recognition analysis of wet gas mixture; a feasibility
study. Can.J.Spectrosc. 27, 127-131.

Hoogerbrugge, R., Willig, S.J., Kistemaker, P.G. (1983).
Discriminant analysis by double stage principal component analysis. Anal.Chem. 55, 1710-1712.

Howery, D.G. (1977). The unique role of target-transform-
ation factor analysis in the chemometric revolution, in
Chemometrics: Theory and Application. ACS Symp. Series
No.52. (Edited by B.R. Kowalski). p.73, Amer.Chem.Soc.,
Washington.

Howery, D.G., Hirsch, R.F. (1983). Chemometrics in the
chemistry curriculum. J.Chem.Educ. 60, 656-659.

Hsu, F.S., Good, B.W., Parrish, M.E., Crews, T.D. (1982).
Pattern recognition for analysis of cigarette smoke by
capillary gas chromatography. 1. Total particulate
matter (TPM). J.High Res.Chromatogr. 5, 648-655.

Huber, J.F.K., Reich, G. (1980). Extraction of information
on the chemical structure of monofunctional compounds
from retention data in gas-liquid chromatography by
pattern recognition methods. Anal.Chim.Acta 122, 139-149.

Huber, J.F.K., Reich, G. (1984). Characterization and
selection of stationary phase for gas-liquid chromatography by pattern recognition. J.Chromatogr. 294, 15-29.

Ichise, M., Yamagishi, H., Oishi, H., Kojima, T. (1980).
Application of pattern recognition techniques to qualitative electrochemical analysis. 3. Data compression
of non-linear static characteristics of the electrode
process. J.Electroanal.Chem. 108, 213-222.

Ichise, M., Yamagishi, H., Kojima, T. (1980a). Analog
feedback linear learning machine applied to the peak
height analysis in stair-case polarography. J.Electro-
anal.Chem. 113, 41-49.

Ichise, M., Yamagishi, H., Oishi, H., Kojima, T. (1980b).
Application of pattern recognition techniques to qualitative electrochemical analysis. 2. On-line measurements
of double-layer capacity by the superimposition method
and adaptive compensation of charging current by the

capacitance multiplier. J.Electroanal.Chem. <u>106</u>, 35-45.

Ichise, M., Yamagishi, H., Oishi, H., Kojima, T. (1982). Applications of pattern recognition techniques to qualitative electrochemical analysis. 4. Simultaneous determination of the linear and non-linear characteristics in the electrode process by the analog feedback linear learning machine. J.Electroanal.,Chem. <u>132</u>, 85-97.

Ichise, M., Kojima, T., Yamagishi, H. (1983). Application of pattern recognition techniques to qualitative electrochemical analysis. 5. Trial retrieval of the electrode system. J.Electroanal.Chem.Interfac. <u>147</u>, 97-105.

IEEE Trans.Comput. Special Issue on Feature Extraction and Selection in Pattern Recognition (1971), <u>20</u>, 967-1117.

Ioffe, I.I., Dobrotvorskii, A.M., Belozerskikh, V.A. (1980). Use of mathematical methods of the theory of pattern recognition for the classification and planning of catalyst searches. Mat.Metody v Khimii. Materialy 3-i Vses. Konf., Rostov, 1979. pp.61-69, Moscow. Chem. Abstr. (1981), <u>95</u>, 157 332.

Ioffe, I.I., Dobrotvorskii, A.M., Belozerskikh, V.A. (1983). Prediction and analysis of the mechanism of the action of heterogeneous catalysts using computer methods of pattern recognition. Usp.Khim. <u>52</u>, 402-425.

Isaszegi-Vass, I., Fuhrmann, G., Horváth, C., Pungor, E., Veress, G.E. (1984). Application of pattern recognition in chromatography, in <u>Modern Trends in Analytical Chemistry</u>, Part B. (Edited by E. Pungor, G.E. Veress and I. Buzás). p.109, Akadémiai Kiadó, Budapest.

Jansen, R.T.P., Pijpers, F.W., de Valk, G.A.J.M. (1981). Application of pattern recognition for discrimination between routine analytical methods used in clinical laboratories. Anal.Chim.Acta. <u>133</u>, 1-18.

Jansen, R.T.P., Pijpers, F.W., de Valk, G.A.J.M. (1981a). A technique for the objective assessment of routine analytical methods in clinical laboratories using pattern recognition. Ann.Clin.Biochem. 18, 218-225.

Jellum, E., Bjørnson, I., Nesbakken, R., Johansson, E., Wold, S. (1981). Classification of human cancer cells by means of capillary gas chromatography and pattern recognition analysis. J.Chromatogr. 217, 231-237.

Jellum, E. (1981a). Computerized gas chromatography- mass spectrometry in biomedical studies. Coordination and handling of the data. CODATA Bull. 41, 8-10.

Jerkovich, G. (1980). Training algorithm for evaluating mass spectra. Kem.Kozl. 54, 349-351. Chem. Abstr. (1981), 95, 71 820.

Johansson, E., Wold, S., Sjödin, K. (1984). Minimizing effects of closure on analytical data. Anal.Chem. 56, 1685-1688.

Johnels, D., Edlund, U., Johansson, E., Wold, S. (1983). A multivariate method for carbon-13 NMR chemical shift predictions using partial least-squares data analysis. J.Magn.Reson. 55, 316-321.

Johnels, D., Edlund, U., Grahn, H., Hellberg, S., Sjöström, M., Wold, S. (1983a). Clustering of aryl carbon-13 nuclear magnetic resonance substituent chemical shifts. A multivariate data analysis using principal components. J.Chem.Soc., Perkin Trans. 2, 863-871.

Jurs, P.C., Isenhour, T.L. (1975). Chemical Applications of Pattern Recognition. Wiley-Interscience, New York.

Jurs, P.C., Chou, J.T., Yuan, M. (1979). Computer assisted structure-activity studies of chemical carcinogens. A heterogeneous data set. J.Med.Chem. 22, 476-483.

Jurs, P.C., Chou, J.T., Yuan, M. (1979a). Studies of chemical structure-biological activity relations using pattern recognition, in Computer Assisted Drug Design. (Edited by E.C. Olsen and R.E. Christoffersen). p.103, Amer.Chem.Soc., Washington.

Jurs, P.C., Ham, C.L., Brugger, W.E. (1981). Computer

assisted studies of chemical structure and olfactory
quality using pattern recognition techniques, in
Odor Quality and Chemical Structure. (Edited by H.R.
Moskowitz and C.B. Warren). p.143, Amer.Chem.Soc.,
Washington.

Jurs, P.C. (1983). Computer assisted studies of structure
-activity relations using pattern recognition. Drug
Inf.J. 17, 219-229.

Jurs, P.C. (1983a). Studies of relationships between molecular and biological activity by pattern recognition
methods, in Structure-Activity Correlation as a
Predictive Tool in Toxicology. (Edited by L. Golberg).
p.93, Hemisphere, New York.

Jurs, P.C., Hasan, M.N., Henry, D.R., Stouch, T.R.,
Whalen-Pedersen, E.K. (1983b). Computer assisted studies
of molecular structure and carcinogenic activity. Fund.
Appl.Tox. 3, 343-349.

Jurs, P.C. (1984). Computer applications in chemistry:
a University course, in Computer Education of Chemistry.
(Edited by P. Lykos). p.1, Wiley, New York.

Jurs, P.C., Stouch, T.R., Czerwinski, M., Narvaez, J.N.
(1985). Computer-assisted studies of molecular structure
-biological activity relationships. J.Chem.Inf.Comput.
Sci. 25, 296-308.

Justice, J.B., Isenhour, T.L. (Editors). (1982). Digital
Computers in Analytical Chemistry. Part 1. Academic
Press, New York.

Justice, J.B., Isenhour, T.L. (Editors). (1982a). Digital
Computers in Analytical Chemistry. Part 2. Academic
Press, New York.

Kaberline, S.L., Wilkins, C.L. (1978). Evaluation of the
super-modified simplex for use in chemical pattern recognition. Anal.Chim.Acta 103, 417-428.

Kateman, G., Pijpers, F.W. (1981). Quality Control in
Analytical Chemistry. p.207, Wiley, New York.

Kaufman, L., Pierreux, A., Rousseeuw, P., Derde, M.P., Detaevernier, M.R., Massart, D.L., Platbrood, G. (1983). Clustering on a microcomputer with an application to the classification of coals. Anal.Chim.Acta 153, 253-260.

Kentgens, A.P.M., Pijpers, F.W., Vertogen, G. (1983). A critical assessment of models predicting alloying behaviour by means of pattern recognition. Anal.Chim. Acta 151, 167-178.

Killeen, T.J., Eastwood, D., Hendrick, M.S. (1981). Oil-matching by using a simple vector model for fluorescence spectra. Talanta 28, 1-6.

Kirschner, G.L., Kowalski, B.R. (1979), in Drug Design. Vol.8. (Edited by E.J. Ariens). p.73, Academic Press, New York.

Kittler, J. (1977). On the discriminant vector method of feature selection. IEEE Trans.Comput. 26, 604-606.

Klee, M.S., Harper, A.M., Rogers, L.B. (1981). Effects of normalization on feature selection in pyrolysis gas chromatography of coal tar pitches. Anal.Chem. 53, 801-805.

Klopman, G. (1984). Artificial intelligence approach to structure-activity studies. Computer automated structure evaluation of biological activity of organic molecules. J.Amer.Chem.Soc. 106, 7315-7321.

Koontz, W.L.G., Narendra, P.M., Fukunaga, K. (1975). A branch and bound clustering algorithm. IEEE Trans. Comput. 24, 908-915.

Kowalski, B.R., Schatzki, T.F., Stross, F.H. (1972). Classification of archaeological artifacts by applying pattern recognition to trace element data. Anal.Chem. 44, 2176-2180.

Kowalski, B.R., Bender, C.F. (1972a). Pattern recognition. A powerful approach to interpreting chemical data. J.Amer.Chem.Soc. 94, 5632-5639.

Kowalski, B.R., Bender, C.F. (1973). Pattern recognition.

2. Linear and nonlinear methods for displaying chemical data. J.Amer.Chem.Soc. 95, 686-693.

Kowalski, B.R. (1975). Chemometrics: Views and propositions. J.Chem.Inf.Comput.Sci. 15, 201-203.

Kowalski, B.R., Bender, C.F. (1976). An orthogonal feature selection method. Pattern Recognition 8, 1-8.

Kowalski, B.R. (Editor). (1977). Chemometrics: Theory and Application. ACS Symp. Series No.52, Amer.Chem.Soc., Washington.

Kowalski, B.R. (1980). Chemometrics. Anal.Chem. 52, 112R-122R.

Kowalski, B.R. (1981), in the article Borman, S.A. New directions in analytical chemistry. Anal.Chem. 53, 703A-706A.

Kowalski, B.R., Wold, S. (1982). Classification, pattern recognition and reduction of dimensionality, in Pattern Recognition in Chemistry. (Edited by P.R. Krishnaiah and L.N. Kanal). North-Holland, Amsterdam.

Kryger, L. (1980). Computational practice in pattern recognition. Anal.Proc. (London) 17, 135-137.

Kryger, L. (1981). Interpretation of analytical chemical information by pattern recognition methods. A survey. Talanta 28, 871-887.

Kuchynka, K., Fusek, J., Štrouf, O. (1981). Modelling of Fischer-Tropsch catalytic synthesis by pattern recognition method. Chemisorption of hydrogen on metals. Coll.Czechoslov.Chem.Commun. 46, 2328-2335.

Kuchynka, K., Fusek, J., Štrouf, O. (1981a). Catalytic activity of transition metals in hydrogenolysis of ethane calculated by the simplex method. Coll.Czechoslov.Chem.Commun. 46, 52-57.

Kuchynka, K., Štrouf, O., Fusek, J. (1982). The pattern recognition approach to catalysis. Proc. Czechoslov. Conf. Prep. Properties Metal. Heterogen. Catalysts, 1st, Bechyně. p.53.

Kvalheim, O.M., Oygard, K., Grahl-Nielsen, O. (1983). SIMCA multivariate data analysis of blue mussel components in environmental pollution studies. Anal.Chim. Acta 150, 145-152.

Kwan, W.-O., Kowalski, B.R., Skogerboe, R.K. (1979). Pattern recognition analysis of elemental data. Wines of Vitis vinifera cv Pinot Noir from France and the United States. J.Agric.Food Chem. 27, 1321-1326.

Kwan, W.-O., Kowalski, B.R. (1980). Correlation of objective chemical measurements and subjective sensory evaluations. Anal.Chim.Acta 122, 215-222.

Kwan, W.-O., Kowalski, B.R. (1980a). Pattern recognition analysis of gas chromatographic data. Geographic classification of wines of Vitis vinifera cv Pinot Noir from France and the United States. J.Agric.Food Chem. 28, 356-359.

Kwan, P.W., Clark, R.C.,Jr. (1981). Assessment of oil contamination in the marine environment by pattern recognition analysis of paraffinic hydrocarbon content of mussels. Anal.Chim.Acta 133, 151-168.

Kwiatkowski, J., Riepe, W. (1984). Optimization of spectral information selection from a spectral library to establish useful training sets in pattern recognition or reliable retrieval methods, in Modern Trends in Analytical Chemistry, Part B. (Edited by E. Pungor, G.E. Veress and I. Buzás). p.75, Akadémiai Kiadó, Budapest.

Lam, R.B., Foulk, S.J., Isenhour, T.L. (1981). Clipped Fourier transform mass spectral compression algorithm for microcomputer-compatible search systems. Anal.Chem. 53, 1679-1684.

Lam, T.F., Wilkins, C.L., Brunner, T.R., Soltzberg, L.J., Kaberline, S.L. (1976). Large-scale mass spectral analysis by simplex pattern recognition. Anal.Chem. 48, 1768-1774.

Läuter, J., Hampicke, J. (1973). MVDA (Mehrdimensionale
Varianz und Diskriminanzanalyse). Berlin-Buch, Zentral-
inst. Herz- und Kreislaufregulationsforsch., Akad.
Wissenschafts.

Lea, R.E., Bramston-Cook, R., Tschida, J. (1983). Pattern
recognition for identification and quantification of
complex mixtures in chromatography. Anal.Chem. 55,
626-629.

Lewi, P.J. (1980), in Drug Design, Vol.10. (Edited by
E.J. Ariens). p.307, Academic Press, New York.

Lindberg, W., Persson, J.-Å., Wold, S. (1983). Partial
least-squares method for spectrofluorimetric analysis
of mixtures of humic acid and ligninsulfonate. Anal.
Chem. 55, 643-648.

Loftsgaarden, D.O., Quesenberry, C.P. (1965). A nonparametric estimate of a multivariate density function.
Ann.Math.Stat. 36, 1049-1051.

Lukovits, I., Lopata, A. (1980). Decomposition of pharmacological activity indices into mutually independent
components using principal component analysis. J.Med.
Chem. 23, 449-459.

Lukovits, I. (1983). Quantitative structure-activity
relationships employing independent quantum chemical
indexes. J.Med.Chem. 26, 1104-1109.

Lundgren, L., Novelius, G., Stenhagen, G. (1981). Selection of biochemical characteristics in the breeding for
pest and disease resistance: A method based on analogy
analysis of chromatographic separation patterns for
emitted plant substances. Hereditas 95, 173-179.

Maclagan, R.G.A.R., Mitchell, M.J. (1980). Pattern recognition and interpretation of nucleoside mass spectra.
Austral.J.Chem. 33, 1401-1408.

Mager, P.P. (1980). Correlation between quantitatively
distributed predicting variables and chemical terms
in acridine derivatives using principal component

analysis. Biom.J. <u>22</u>, 813-825.

Mager, P.P. (1981). Pattern recognition and time-dependent QSAR applied to morphinomimetic opioids. Act.Nerv.Super. <u>23</u>, 136-156. Chem. Abstr. (1981), <u>95</u>, 125 881.

Mahle, N.H., Ashley, J.W. (1979). Application of a correlation coefficient pattern recognition technique to low resolution mass spectra. Comput.Chem. <u>3</u>, 19-23.

Malinowski, E.R., Howery, D.G. (1980). <u>Factor Analysis in Chemistry</u>. Wiley, New York.

Malissa, H. (1984). Analytical chemistry today and tomorrow. Fresenius Z.Anal.Chem. <u>319</u>, 357-363.

Marchese, F.T., Beveridge, D.L. (1984). Pattern recognition approach to the analysis of geometrical features of solvation: application to the aqueous hydration of Li^+, Na^+, K^+, F^-, and Cl^-. J.Amer.Chem.Soc. <u>106</u>, 3713-3720.

Marshall, R.J., Turner, R., Yu, H., Cooper, E.H. (1984). Cluster analysis of chromatographic profiles of urine proteins. J.Chromatogr. <u>297</u>, 235-244.

Martens, M., Martens, H., Wold, S. (1983). Preference of cauliflower related to sensory descriptive variables by partial least squares (PLS) regression. J.Sci.Food Agric. <u>34</u>, 715-724.

Martin, Y.C., Panas, H.N. (1979). Mathematical considerations in series design. J.Med.Chem. <u>22</u>, 784-791.

Martinsen, D.P. (1981). Survey of computer-aided methods for mass spectral interpretation. Appl.Spectrosc. <u>35</u>, 255-266.

Massart, D.L., Michotte, Y. (1979). Use of pattern recognition methods in analytical chemistry. Bull.Soc.Chim. Fr. (Pt.1), 293-300.

Massart, D.L., Kaufman, L., Coomans, D. (1980). An operational research model for pattern recognition. Anal.Chim.Acta <u>122</u>, 347-355.

Massart, D.L., Kaufman, L., Coomans, D., Esbensen, K.H.

175

(1981). The classification of iron meteorites. A re-
appraisal of existing classification. Bull.Soc.Chim.
Belg. 90, 281-288.

Massart, D.L. (1981a). Clustering for microanalytical
data. Nat. Aim Methods Microchem. Proc. Int. Microchem. Symp., 8th, 1980. (Edited by H. Malissa,
M. Grasserbauer and R. Belcher). p.275, Springer,
Vienna.

Massart, D.L. (1982). Extraction of information from large
data sets by pattern recognition. Fresenius Z.Anal.Chem.
311, 318.

Massart, D.L., Buydens, L., Armanino, C., Broeckaert, I.,
Coomans, D., Dekker, W.H., Derde, M.P., Detaevernier,
M., Esbensen, K.H., Forina, M., Jonckheer, M., Kaufman,
L. (1982a). Application of pattern recognition.
Fresenius Z.Anal.Chem. 311, 448.

Massart, D.L., Kaufman, L., Esbensen, K.H. (1982b).
Hierarchical nonhierarchical clustering strategy and
application to classification of iron meteorites according to their trace element patterns. Anal.Chem. 54,
911-917.

Massart, D.L., Plastria, F., Kaufman, L. (1983). Non-
hierarchical clustering with MASLOC. Pattern Recognition
16, 507-516.

Massart, D.L., Kaufman, L. (1983a). The Interpretation
of Analytical Chemical Data by the Use of Cluster
Analysis. Wiley, New York.

Massart-Leën, A.-M., Massart, D.L. (1981). The use of
clustering techniques in the elucidation or confirmation
of metabolic pathways. Biochem.J. 196, 611-618.

Matsubara, M. (1982). Comparison of composition and
structural parameters of heavy oils by statistical
methods. Pan-Pac. Synfuels Conf. 2, 539-546. Chem.
Abstr. (1984), 101, 40 697.

McConnell, M.L., Rhodes, G., Watson, U., Novotny, M.

(1979). Application of pattern recognition and feature extraction techniques to volatile constituent metabolic profiles obtained by capillary gas chromatography. J.Chromatogr. 162, 495-506.

McCown, S.M., Manos, C.G.,Jr., Pitzer, D.R., Earnest, C.M. (1982). The (r,Q) matrices. A tool for the manipulation of chromatographic patterns. Analyst (London) 107, 1393-1406.

Mc Lafferty, F.W., Venkataraghavan, R. (1979). Computer techniques for mass spectral identification. J.Chromatogr. Sci. 17, 24-29.

Meglen, R.R., Erickson, G.A. (1983). Application of pattern recognition to the evaluation of contamination from oil shale retorting. Environ. Solid Wastes: Charact., Treat., Disposal, 4th, 1981. p.369. Chem. Abstr. (1984), 100, 197 466.

Meier, B.U., Bodenhausen, G., Ernst, R.R. (1984). Pattern recognition in two-dimensional NMR spectra. J.Magn. Resonance 60, 161-163.

Meisel, W.S., Jolley, M., Heller, S.R., Milne, G.W.A. (1979). The role of pattern recognition in the computer -aided classification of mass spectra. Anal.Chim.Acta 112, 407-416.

Meister, A. (1984). Estimation of component spectra by the principal components method. Anal.Chim.Acta 161, 149-161.

Meites, L. (1982). The limit of detection of a weak acid, in the presence of another, by potentiometric acid-base titrimetry and deviation-pattern recognition. Anal. Lett. 15, 507-517.

Meites, L. (1982a). The limit of detection of a reactant, in the presence of another, by kinetic analysis and deviation-pattern recognition. Anal.Lett. 15, 1149-1158.

Mellon, F.A. (1979), in Mass Spectrometry, Vol.5. (Edited by R.A.W. Johnstone). p.100, The Chem.Soc., London.

Meuzelaar, H.L.C. (1982). Characterization of Rocky
Mountain coals and coal liquids by computerized analytical techniques. Report, DOE/PC/30242-T4. Chem. Abstr.
(1983), 99, 56 106.

Milina, R., Dimov, N., Dimitrova, M. (1983). Classification system for recognition of low-molecular-weight
substances from their pyrograms. Chromatographia 17,
29-32.

Miyashita, Y., Seki, T., Takahashi, Y., Daiba, S., Tanaka,
Y., Yotsui, Y., Abe, H., Sasaki, S. (1981). Computer-
assisted structure-carcinogenicity studies on polycyclic aromatic hydrocarbons by pattern recognition
methods. Anal.Chim.Acta 133, 603-613.

Miyashita, Y., Takahashi, Y., Yotsui, Y., Abe, H., Sasaki,
S. (1981a). Application of pattern recognition to
structure-activity problems. Use of minimal spanning
tree. Anal.Chim.Acta 133, 615-624.

Miyashita, Y., Takahashi, Y., Yotsui, Y., Abe, H., Sasaki.
S. (1981b). The use of cluster analysis and display
method of pattern recognition in structure-activity
studies of antibiotics. CODATA Bull. No.41, 37-41.

Miyashita, Y., Abe, H., Sasaki, S. (1981c). The pattern
recognition explaining the carcinogenic activity-
structure relationship. Kagaku (Kyoto) 36, 351-356.
Chem. Abstr. (1981), 95, 91 710.

Miyashita, Y., Takahashi, Y., Abe, H., Sasaki, S. (1982).
Structure-activity correlation by pattern recognition.
Kagaku no Ryoiki, Zokan, 253-276. Chem. Abstr. (1982),
97, 207 543.

Miyashita, Y., Takahashi, Y., Dauba, S.-I., Abe, H.,
Sasaki, S. (1982a). Computer-assisted structure-
carcinogenicity studies on polynuclear aromatic hydrocarbons by pattern recognition methods. The role of
the bay and L-regions. Anal.Chim.Acta 143, 35-44.

Molnár, P., Liszonyi, G.M., Orsi, F. (1984). Some results

of pattern recognition and cluster analysis methods in
sensory evaluation of food quality, in <u>Modern Trends</u>
<u>in Analytical Chemistry</u>, Part B. (Edited by E. Pungor,
G.E. Veress and I. Buzás). p.137, Akadémiai Kiadó,
Budapest.

Morgan, S.I. (1981). Structure correlation and pattern
recognition in analytical pyrolysis. Org.Coat.Plast.
Chem. <u>44</u>, 600-603. Chem. Abstr. (1983), <u>98</u>, 64 867.

Moriguchi, I., Komatsu, K., Matsishita, Y. (1980).
Adaptive least-squares method applied to structure-
activity correlations of hypotensive N-alkyl-N''-cyano-
N'-pyridyl-quanidines. J.Med.Chem. <u>23</u>, 20-26.

Moriguchi, I., Komatsu, K., Matsushita, Y. (1981). Pattern
recognition for the study of structure-activity relationships. Uses of the adaptive least-squares method and
linear discriminant analysis. Anal.Chim.Acta <u>133</u>,
625-636.

Moriguchi, I., Komatsu, K. (1981a). Structure-activity
studies of withaferin analogues using the adaptive least
-squares method. Eur.J.Med.Chem., Chim.Therapeutica
<u>16</u>, 19-23.

Musch, G., DeSmet, M., Massart, D.L. (1985). Expert system
for pharmaceutical analysis. 1. Selection of the detection system in high-performance liquid chromatographic
analysis: UV versus amperometric detection. J.Chromatogr.
<u>348</u>, 97-110.

Musumarra, G., Wold, S., Gronowitz, S. (1981). Application
of principal component analysis to C-13 NMR shifts of
chalcones and their thiophene and furan analogues:
a useful tool for the shift assignment and for the study
of substituent effects. Org.Magn.Resonance <u>17</u>, 118-123.

Musumarra, G., Scarlata, G., Wold, S. (1981a). Studies of
substituent effects by carbon-13 NMR spectroscopy. 4.
Application of principal components analysis to 13-C NMR
shifts of (Z)-[alfa-(p-substituted-phenyl)-beta-(5-sub-

stituted-2-thienyl)acrylonitrile]. Gazz.Chim.Ital. <u>111</u>, 499-502.

Nauer, G., Kny, E., Haevernick, T.E. (1980). Chemical composition of Roman glass finds as an aid to their numerical classification. 4. Numerical classification of 161 glasses from Regensburg. Glastechn.Ber. <u>53</u>, 232-236.

Nestrick, T.J., Lamparski, L.L., Townsend, D.I. (1980). Identification of tetrachlorodibenzo-p-dioxin isomers at the 1-ng level by photolytic degradation and pattern recognition techniques. Anal.Chem. <u>52</u>, 1865-1874.

Nie, N.N., Hull, C.H., Jenkins, J.G., Steinbrenner, K., Bent, D. (1975). Statistical Package for the Social Sciences (SPSS). McGraw-Hill, New York.

Nilsson, N.J. (1965). Learning Machines. McGraw-Hill, New York.

Nordén, B., Edlund, U., Johnels, D., Wold, S. (1983). Simplified C-13 NMR parameters related to the carcinogenic potency of polycyclic aromatic hydrocarbons. QSAR <u>2</u>, 73-76.

Ogino, A., Matsumura, S., Fujita, T. (1980). Structure-activity study of antiulcerous and antiinflammatory drugs by discriminant analysis. J.Med.Chem. <u>23</u>, 437-444.

Ordukhanyan, A.A., Sarkisyan, A.S., Landau, M.A., Mndzhoyan, Sh.L., Ter-Zakharyan, Yu.Z. (1983). Classification of chemical compounds by their spectrum of biological activity. 1. Method.Khim.-Farm.Zh. <u>17</u>, 63-66.

Owens, P.M., Lam,R.B., Isenhour, T.L. (1982). Factor analysis for real-time gas chromatography/Fourier transform infrared spectrometric chromatogram reconstruction. Anal.Chem. <u>54</u>, 2344-2347.

Owens, P.M., Isenhour, T.L. (1983). Infrared spectral compression procedure for resolution independent search systems. Anal.Chem. <u>54</u>, 1548-1553.

Parrish, M.E., Good, B.W., Hsu, F.S., Hatch, F.W.,

Ennis, D.M., Douglas, D.R., Shelton, J.H., Watson, D.C., Reiley, C.N. (1981). Computer-enhanced high-resolution gas chromatography for the discriminative analysis of tabacco smoke. Anal.Chem. 53, 826-831.

Parrish, M.E., Good, B.W., Jeltema, M.A., Hsu, F.S. (1983). Pattern recognition and capillary gas chromatography in the analysis of the organic gas phase of cigarette smoke. Anal.Chim.Acta 150, 163-170.

Peredunova, I.V., Kruglyak, Y.A. (1983). A program for establishing the relationship between the structure of chemical compounds and their properties (Logical-structural scheme of pattern recognition). Zh.Strukt. Khim. 24, 166-167.

Perone, S.P., Kryger, L., Byers, W.A. (1982). Interpretation of electroanalytical data using computerized pattern recognition. Report, TR-29; Order No. AD-A121504. Chem. Abstr. (1983), 99, 47 291.

Perone, S.P., Spindler, W.C. (1984). Battery lifetime prediction by pattern recognition. Application to lead -acid battery life-cycling test data. J.Power Sources 13, 23-38.

Petit, B., Potenzone, R.,Jr., Hopfinger, A.J., Klopman, G., Shapiro, M. (1979), in Computer-Assisted Drug Design. ACS Symp. Series No.112. (Edited by E.C. Olson and R.E. Christoffersen). p.553, Amer.Chem.Soc., Washington.

Pijpers, F.W., VanGaal, H.L.M., Van der Linden, J.G.H. (1979). Qualitative classification of dithiocarbamate compounds from carbon-13 NMR and IR spectroscopic data by pattern recognition techniques. Anal.Chim.Acta 112, 199-209.

Pijpers, F.W., Vertogen, G. (1982). Can superconductivity be predicted with the aid of pattern recognition technique? J.Phys., Paris 43, 97-106.

Pijpers, F.W. (1984). Failures and successes with pattern recognition for solving problems in analytical chemistry.

Analyst <u>109</u>, 299-303.

Pillay, A.E., Peisach, M. (1981). Human hair analysis by PIXE: Pattern recognition of trace element composition. J.Radioanal.Chem. <u>63</u>, 85-95.

Pino, J.A., McMurry, J.E., Jurs, P.C., Lavine, B.K., Harper, A.M. (1985). Application of pyrolysis-gas chromatography-pattern recognition to the detection of cystic fibrosis heterozygotes. Anal.Chem. <u>57</u>, 295-302.

Polak, E. (1971). Computational Methods in Optimization. Academic Press, New York.

Pungor, E., Veress, G.E., Buzás, I. (Editors). (1984). Pattern recognition in analytical chemistry, in Modern Trends in Analytical Chemistry, Part B. Anal. Chem. Symp. Series, Vol.18. Elsevier, Amsterdam, and Akadémiai Kiadó, Budapest.

Randić, M., Wilkins, C.L. (1979). Graph theoretical approach to recognition of structural similarity in molecule. J.Chem.Inf.Comput.Sci. <u>19</u>, 31-37.

Rasmussen, G.T., Hohne, B.A., Wieboldt, R.C., Isenhour, T.L. (1979). Identification of components in mixture by a mathematical analysis of mass spectral data. Anal. Chim.Acta <u>112</u>, 151-164.

Rasmussen, G.T., Ritter, G.L., Lowry, S.R., Isenhour, T.L. (1979a). Fisher discriminant functions for a multilevel mass spectral filter network. J.Chem.Inf.Comput.Sci. <u>19</u>, 255-259.

Rasmussen, G.T., Isenhour, T.L. (1979b). Library retrieval of infrared spectra based on detailed intensity information. Appl.Spectrosc. <u>33</u>, 371-376.

Rasmussen, G.T., Isenhour, T.L., Marshall, J.C. (1979c). Mass spectral library searches using ion series data compression. J.Chem.Inf.Comput.Sci. <u>19</u>, 98-104.

Rasmussen, G.T., Isenhour, T.L. (1979d). The evaluation of mass spectral search algorithms. J.Chem.Inf.Comput. Sci. <u>19</u>, 179-186.

Richards, J.A., Griffiths, A.G. (1979). On confidence in the results of learning machines trained on mass spectra. Anal.Chem. 51, 1358-1361.

Ritter, G.L., Lowry, S.R., Wilkins, C.L., Isenhour, T.L. (1975). Simplex pattern recognition. Anal.Chem. 47, 1951-1956.

Rose, S.L., Jurs, P.C. (1982). Computer-assisted studies of structure-activity relationships of N-nitroso compounds using pattern recognition. J.Med.Chem. 25, 769-776.

Rosenblatt, F. (1962). Principle of Neurodynamics: Perceptrons and the Theory of Brain Mechanisms. Spartan Books, Washington.

Rosenblatt, F. (1964). A model for experimental storage in neural networks, in Computer and Information Sciences. (Edited by J.T. Tou and R.H. Wilcox). p.16, Spartan Books, Washington.

Rotter, H., Varmuza, K. (1975). Criteria for the evaluation of classifiers for the automatic interpretation of spectra (Pattern recognition). Org.Mass Spectrometry 10, 874-884.

Rusling, J.F. (1983). Computerized method for mechanistic classification of one-electron potentiostatic current -potential curves. Anal.Chem. 55, 1713-1718.

Rusling, J.F. (1983a). Applications of a computerized method for mechanistic classification of one-electron potentiostatic current-potential curves. Anal.Chem. 55, 1719-1723.

Rusling, J.F. (1984). Computerized interpretation of electrochemical data using deviation-pattern recognition. Trends Anal.Chem. 3, 91-94.

Ryabova, M.S., Sautin, S.N., Volin, Yu.M., Lazarev, S.Ya., Shibaev, V.A. (1983). Use of pattern recognition in investigation of epoxidation of ethylene-propylene rubber by performic acid. Zh.Prikl.Khim. 56, 1116-1122.

Ryan, P.B., Barr, R.L., Todd, H.D. (1980). Simplex techniques for nonlinear optimization. Anal.Chem. <u>52</u>, 1460-1467.

Saarinen, L. (1983). Gas chromatographic analysis of multi-component mixtures by using a microcomputer. Kem.-Kemi <u>10</u>, 814-815. Chem. Abstr. (1984), <u>100</u>, 167 413.

Saltiel, J., Eaker, D.W. (1984). Principal component analysis to 1-phenyl-2-(2-naphthyl)ethene fluorescence. Four components not two. J.Amer.Chem.Soc. <u>106</u>, 7624-7626.

Sammon, J.W.,Jr. (1969). A nonlinear mapping for data structure analysis. IEEE Trans.Comput. <u>18</u>, 401-409.

Scarminio, I.S., Bruns, R.E., Zagatto, E.A.G. (1982). Pattern recognition classification of mineral waters based on spectrochemical analysis. Energ.Nucl.Agric. <u>4</u>, 99-111.

Schachterle, S.D. (1980). Classification of voltammetric data using computerized pattern recognition. Diss. Abstr. Int. B <u>42</u>, 196.

Schachterle, S.D., Perone, S.P. (1981). Classification of voltammetric data by computerized pattern recognition. Anal.Chem. <u>53</u>, 1672-1678.

Scoble, H.A., Fasching, J.L., Brown, P.R. (1983). Chemometrics and liquid chromatography in the study of acute lymphocytic leukemia. Anal.Chim.Acta <u>150</u>, 171-181.

Sepaniak, M.J., Yeung, E.S. (1981). Coal classification by HPLC and three-dimensional detection. Prepr. Pap.-Am.Chem.Soc., Div.Fuel Chem. <u>26</u>, 1-6. Chem. Abstr. (1982), <u>97</u>, 130 367.

Sjöström, M., Kowalski, B.R. (1979). A comparison of five pattern recognition methods based on the classification results from six real data bases. Anal.Chim.Acta <u>112</u>, 11-30.

Sjöström, M., Wold, S. (1980). SIMCA: a pattern recognition method based on principal component models, in

184

Pattern Recognition in Practice. (Edited by E.S. Gelsema and L.N. Kanal). p.351, North-Holland, Amsterdam.

Sjöström, M., Wold, S., Lindberg, W., Persson, J.-A., Martens, H. (1983). A multivariate calibration problem in analytical chemistry solved by partial least-squares models in latent variables. Anal.Chim.Acta 150, 61-70.

Smith, A.B.,III, Belcher, A.M., Epple, G., Jurs, P.C., Lavine, B.K. (1985). Computerized pattern recognition: a new technique for the analysis of chemical communication. Science 228 (4696), 175-177.

Small, G.W., Rasmussen, G.T., Isenhour, T.L. (1979). An infrared search system based on direct comparison of interferograms. Appl.Spectrosc. 33, 444-450.

Snygg, B.G., Andersson, J.E., Krall, C.A., Stölbran, U.M., Akesson, C.A. (1979). Separation of botulinum-positive and negative fish samples by means of a pattern recognition method applied to headspace gas chromatograms. Appl.Env.IR.Microbiol. 38, 1081-1085. From Isaszegi-Vass.

Söderström, B., Wold, S., Blomquist, G. (1982). Pyrolysis -gas chromatography combined with SIMCA pattern recognition for classification of fruit-bodies of some Ectomycorrhizal Suillus species. J.Gen.Microbiol. 128, 1783-1784.

Sogliero, G., Eastwood, D., Ehmer, R. (1982). Some pattern recognition considerations for low-temperature luminiscence and room temperature fluorescence spectra. Appl.Spectrosc. 36, 110-116.

Solominova, T.S., Maksimov, G.G., Semenov, V.A. (1984). Predicting acute toxicity of organic compounds by the pattern recognition method. Khim.-Farm.Zh. 18, 181-188.

Sparks, D.T., Lam, R.B., Isenhour, T.L. (1982). Quantitative gas chromatography/Fourier transform infrared spectrometry with integrated Gram-Schmidt reconstruction intensities. Anal.Chem. 54, 1922-1926.

Stauffer, D.B., McLafferty, F.W., Ellis, R.D., Peterson, D.W. (1985). Adding forward searching capabilities to a reverse search algorithm for unknown mass spectra. Anal.Chem. 57, 771-773.

Stepanenko, V.E. (1982). Chromatographic identification of individual compounds by pattern recognition. Zh. Anal.Khim. 37, 2230-2234.

Stouch, T.R., Jurs, P.C. (1985). Monte Carlo studies of the classification made by nonparametric linear discriminant functions. J.Chem.Inf.Comput.Sci. 25, 45-50.

Stouch, T.R., Jurs, P.C. (1985a). Monte Carlo studies of the classification made by nonparametric linear discriminant functions. 2. Effects of nonideal data. J.Chem.Inf. Comput.Sci. 25, 92-98.

Stouch, T.R., Jurs, P.C. (1985b). Computer-assisted studies of molecular structure and genotoxic activity using pattern recognition techniques. Environ.Health Persp., 61,329-343.

Streich, W.J., Dove, S., Franke, R. (1980). On the rational selection of test series. 1. Principal component method combined with multidimensional mapping. J.Med. Chem. 23, 1452-1456.

Štrouf, O., Wold, S. (1977). Pattern-recognition search for the basic regularities in the stability of complex hydrides. 1. A simplified model. Acta Chem.Scand. A31, 391-401.

Štrouf, O., Fusek, J. (1979). Model intrinsic dimensionality in pattern recognition analysis of structural data of complex hydrides. Coll.Czechoslov.Chem.Commun. 44, 1370-1378.

Štrouf, O. (1980). Modelling of chemical systems by pattern recognition approach. Czech-Pol. Colloq. Chem. Thermodyn. Phys. Org. Chem., 2nd, Lectures, p.103-120.

Štrouf, O., Fusek, J. (1981). Simplex search for mathematical representation of chemical class structure.

Coll.Czechoslov.Chem.Commun. 46, 58-64.

Štrouf, O., Fusek, J., Kuchynka, K. (1981a). Modelling of catalytic activity of transition metals in hydrogenolysis of ethane by the pattern recognition approach. Coll.Czechoslov.Chem.Commun. 46, 65-71.

Štrouf, O., Kuchynka, K., Fusek, J. (1981b). Modelling of the catalytic Fischer-Tropsch synthesis by pattern recognition. Chemisorption and dissociation of carbon monoxide on metals. Coll.Czechoslov.Chem.Commun. 46, 2336-2344.

Štrouf, O., Fusek, J., Kuchynka, K. (1982). Approximation of the adsorption heats of hydrogen on transition metals by means of an empirical model. Coll.Czechoslov.Chem. Commun. 47, 2363-2367.

Stuper, A.J., Jurs, P.C. (1976). ADAPT: a computer system for automated data analysis using pattern recognition techniques. J.Chem.Inf.Comput.Sci., 16, 99-105.

Stuper, A.J., Brugger, W.E., Jurs, P.C. (1977). Computer system for structure-activity studies using chemical structure information handling and pattern recognition techniques, in Chemometrics: Theory and Application. ACS Symp. Series No.52. (Edited by B.R. Kowalski). p.165, Amer.Chem.Soc., Washington.

Stuper, A.J., Brugger, W.E., Jurs, P.C. (1979). Computer Assisted Studies of Chemical Structure and Biological Function. Wiley, New York.

Takahashi, Y., Miyashita, Y., Abe, H., Sasaki, S., Yotsui, Y., Sano, M. (1980). A structure-biological activity study based on cluster analysis and the nonlinear mapping method of pattern recognition. Anal.Chim.Acta 122, 241-247.

Takahashi, Y., Miyashita, Y., Yotsui, Y., Abe, H., Sasaki, S. (1980a). Application of the simplex method to the classification of pharmacologically active compounds. Joho Kagaku Toronkai Ronbunshu, 3rd, p.13. Chem. Abstr.

(1982), 97, 49 313.

Takahashi, Y., Miyashita, Y., Tanaka, Y., Abe, H., Sasaki, S. (1982). A consideration for structure-taste correlation of perrilartines using pattern-recognition techniques. J.Med.Chem. 25, 1245-1248.

Ten Noever de Braun, M.C., Tas, A.C., van der Greef, J., Bouwman, J. (1984). Pattern recognition in mass spectrometry. Tijdschr.Ned.Ver.Klin.Chem. 9, 11-16. Chem. Abstr. (1984), 100, 205 869.

Tinker, J. (1981). Relating mutagenicity to chemical structure. J.Chem.Inf.Comput.Sci. 21, 3-7.

Tinker, J.F. (1981a). A computerized structure-activity correlation program for relating bacterial mutagenesis activity to chemical structure. J.Comput.Chem. 2, 231-243.

Tou, J.T., Gonzales, R.C. (1974). Pattern Recognition Principles. Addison-Wesley, London. Russian translation, Mir, Moscow, 1978.

Toussaint, G.T. (1974). Bibliography on estimation of misclassification. IEEE Trans.Inform.Theory 20, 472-479.

Tsao, R., Switzer, W.L. (1982). Classification of mono-substituted phenyl rings by parametric methods using infrared and Raman peak heights. Anal.Chim.Acta 134, 111-118.

Tsao, R., Switzer, W.L. (1982a). Classification of mono-substituted phenyl rings by branching tree methods using infrared and Raman peak heights. Anal.Chim.Acta 136, 3-13.

Tsao, R. (1982b). Parametric methods applied to the classification of monosubstituted phenyl rings using infrared and Raman heights. Diss. Abstr. Int.B 43, 130.

Tsao, R., Voorhees, K.J. (1984). Analysis of smoke aerosols from nonflaming combustion by pyrolysis mass spectrometry with pattern recognition. Anal.Chem. 56, 368-373.

Van der Greef, J., Tas, A.C., Bouwman, J., Ten Noever de
Braun, M.C., Schreurs, W.H.P. (1983). Evaluation of
field-desorption and fast atom-bombardment mass spectrometric profiles by pattern recognition techniques. Anal.
Chim.Acta 150, 45-52.

Van der Voet, H., Doornbos, D.A., Meems, M., van de Haar,
G. (1984). The use of pattern recognition techniques in
chemical differentiation between Bordeaux and Bourgogne
wines. Anal.Chim.Acta 159, 159-171.

Van der Voet, H., Doornbos, D.A. (1984a). The improvement
of SIMCA classification by using kernel density estimation. 1. A new probabilistic classification technique
and how to evaluate such a technique. Anal.Chim.Acta
161, 115-123.

Van der Voet, H., Doornbos, D.A. (1984b). The improvement
of SIMCA classification by using kernel density estimation of SIMCA, ALLOC and CLASSY on three data sets.
Anal.Chim.Acta 161, 125-134.

Van Gaal,H.L.M., Diesveld, J.W., Pijpers, F.W., van der
Linden, J.G.M. (1979). ^{13}C NMR spectra of dithiocarbamates. Chemical shifts, carbon-nitrogen stretching
vibration frequencies and π-bonding in the NCS_2 fragment.
Inorg.Chem. 11, 3251-3260.

Varmuza, K., Rotter, H. (1976). Evaluation of automatic
methods of spectra interpretation. Monatsh.Chem. 107,
547-555.

Varmuza, K., Rotter, H. (1978). Judgment of automatic
spectra interpretation methods (Pattern recognition).
Advan.Mass Spectrom. 7, 1099-1102.

Varmuza, K. (1979). Pattern recognition applications and
evaluation of classification. Vestnik Slovenskega
Kemijskega Društva 26, 61-63.

Varmuza, K. (1980). Pattern Recognition in Chemistry.
Springer-Verlag, Berlin.

Varmuza, K. (1980a). Pattern recognition in analytical

chemistry. Anal.Chim.Acta 122, 227-240.

Varmuza, K., Rotter, H. (1980b). Advantages and evaluation of binary classifiers with continuous response (mass spectra interpretation by pattern recognition). Advan.Mass Spectrom. 8, 1541-1548.

Varmuza, K. (1981). Pattern recognition in chemistry. A list of references. Technische Univ. Vienna.

Varmuza, K. (1982). Pattern recognition methods in chemistry. Oesterr.Chem.Z. 83, 117-121.

Varmuza, K. (1983). Some aspects of the application of pattern recognition methods in chemistry, in Computer Applications in Chemistry. (Edited by S.R. Heller and R. Potenzone,Jr.). p.19, Elsevier, Amsterdam.

Varmuza, K., Lohninger, H. (1984). Automatic recognition of chemical classes in series of mass spectra. Euroanalysis 5th, Cracow. Abstr., p.448. (Edited by L. Górski).

Varmuza, K. (1984a). Chemometrie, in Computer in der Chemie. (Edited by E. Ziegler). p.131, Springer-Verlag, Berlin.

Varmuza, K. (1984b). Some aspects of pattern recognition applications in mass spectrometry, in Modern Trends in Analytical Chemistry, Part B. (Edited by E. Pungor, G.E. Veress and I. Buzás). p.99, Akadémiai Kiadó, Budapest.

Veress, G.E. (1982). Pattern recognition in analytical chemistry. Trends Anal.Chem. 1, 374-377.

Veress, G.E., Pungor, E. (1984). The role of pattern recognition in analytical chemical signal interpretation, in Modern Trends in Analytical Chemistry, Part B. (Edited by E. Pungor, G.E. Veress and I. Buzás). p.125, Akadémiai Kiadó, Budapest.

Verhagen, C.J.D.M. (1975). Some general remarks about pattern recognition; its definition, its relation with other disciplines, a literature survey. Pattern

Recognition 7, 109-116.

Verhagen, C.J.D.M., Duin, R.P.W., Groen, F.C.A., Joosten, J.C., Verbeek, P.W. (1980). Progress report on pattern recognition. Rep.Prog.Phys. 43, 785-831.

Visser, T., van der Maas, J.H. (1981). Systematic computer -aided interpretation of vibrational spectra. Anal.Chim. Acta 133, 451-456.

Vorhees, K.J., Durfee, S.L., Bladwin, R.M. (1983). Lique-faction reactivity correlation using pyrolysis mass spectrometry pattern recognition procedures, in Polymer Characterization (Spectroscopic, Chromatographic and Physical Instrumental Methods). (Edited by C.D. Craver). p.677, Amer.Chem.Soc., Washington.

Vuchev, V. (1983). Numerical classification procedures in pattern recognition in organic geochemistry. Sci.Terre, Ser.Inf.Geol. 16, 19-57.

Wang, C.P., Sparks, D.T., Williams, S.S., Isenhour, T.L. (1984). Comparison of methods for reconstructing chromatographic data from liquid chromatography/Fourier transform infrared spectrometry. Anal.Chem. 56, 1268-1272.

Watterson, J.I.W., Sellschop, J.P.F., Erasmus, C., Hart, R.J. (1983). The combination of multielement neutron activation analysis and multivariate statistics for characterization in geochemistry. Int.J.Appl.Radiot. Isot. 34, 407-416.

Watkin, D.J. (1980). Pattern recognition in molecules from a general linear transformation. Acta Crystallogr. 36, 975-978.

Wegscheider, W., Leyden, D.E. (1979). "PAREDS" - an inter-active on-line system for the interpretation of EDXRF data. Advan.X-Ray Anal. 22, 357-368.

Whalen-Pedersen, E.K., Jurs, P.C. (1979). The probability of dichotomization by a binary linear classifier as a function of training set population distribution.

J.Chem.Inf.Comput.Sci. 19, 264-266.

Whalen-Pedersen, E.K., Jurs, P.C. (1981). Computer assisted structure-activity studies of polycyclic aromatic hydrocarbons, in Chemical Analysis and Biological Fate: Polynuclear Aromatic Hydrocarbons. (Edited by M. Cooke and A.J. Dennis). p.55, Battelle Press, Columbus.

Wieboldt, R.C., Hohne, B.A., Isenhour, T.L. (1980). Functional group analysis of interferometric data from gas chromatography/Fourier transform infrared spectroscopy. Appl.Spectrosc. 34, 7-14.

Wienke, D., Danzer, K. (1985). Determination of quality of classification- and cluster-procedures by means of geometrical and information-theoretical criteria, personal communication.

Willett, P. (1982). The calculation of intermolecular similarity coefficients using an inverted file algorithm. Anal.Chim.Acta 138, 339-342.

Willett, P. (1982a). A comparison of some hierarchical agglomerative clustering algorithms for structure-property correlation. Anal.Chim.Acta 136, 29-37.

Willett, P. (1983). Some heuristics for nearest-neighbour searching in chemical structure files. J.Chem.Inf.Comput. Sci. 23, 22-25.

Williams, S.S., Lam, R.B., Isenhour, T.L. (1983). Search system for infrared and mass spectra by factor analysis and eigenvector projection. Anal.Chem. 55, 1117-1121.

Windig, W., Haverkamp, J., Kistemaker, P.G. (1983). Interpretation of sets of pyrolysis mass spectra by discriminant analysis and graphical rotation. Anal.Chem. 55, 81-88.

Wold, H. (1981). Soft modelling: the basic design and some extensions, in Systems Under Indirect Observation. (Edited by K.G. Jöreskog and H. Wold). North-Holland, Amsterdam.

Wold, S. (1974). Pattern cognition and recognition

192

(cluster analysis) based on disjoint principal components models. Techn. Report No.357, Univ. Wisconsin.

Wold, S. (1976). Pattern recognition by means of disjoint
principal component models. Pattern Recognition $\underline{8}$,
127-139.

Wold, S., Sjöström, M. (1977). SIMCA - method for analyzing
chemical data in terms of similarity and analogy, in
Chemometrics: Theory and Application. ACS Symp. Series
No.52. (Edited by B.R. Kowalski). p.243, Am.Chem.Soc.,
Washington.

Wold, S. (1978). Cross validatory estimation of the number
of components in factor and principal components models.
Technometrics $\underline{20}$, 397-406.

Wold, S., Štrouf, O. (1979). Pattern recognition search
for basic regularities in the stability of complex
hydrides. 2. Unsubstituted complexes ABH_4. Acta Chem.
Scand. A$\underline{33}$, 463-467.

Wold, S., Štrouf, O. (1979a). Pattern recognition search
for basic regularities in the stability of complex
hydrides. 3. Monosubstituted complexes ABH_3D. Acta Chem.
Scand. A$\underline{33}$, 521-529.

Wold, S., Johansson, E., Jellum, E., Björnson, I.,
Nesbakken, R. (1981). Application of SIMCA multivariate
data analysis to the classification of gas chromatographic profiles of human brain tissues. Anal.Chim.Acta
$\underline{133}$, 251-259.

Wold, S., Dunn, W.J.,III, Hellberg, S. (1982). Survey of
applications of pattern recognition to structure-activity
problems. Tech. Report 1-1982, Umeå University.

Wold, S., Dunn, W.J.,III. (1983). Multivariate quantitative
structure-activity relationships (QSAR): Conditions for
their applicability. J.Chem.Inf.Comput.Sci. $\underline{23}$, 6-13.

Wold, S., Hellberg, S., Dunn, W.J.,III. (1983a). Computer
methods for the assessment of toxicity. Acta Pharm.
Toxicol. $\underline{52}$, Suppl.II, 158-189.

Wold, S., Albano, C., Dunn, W.J.,III, Esbensen, K., Hellberg, S., Johansson, E., Sjöström, M. (1983b). Pattern recognition: finding and using regularities in multivariate data. Proc. IUPOST Conf. Food Research and Data Analysis. (Edited by H. Martens and H. Russwurm, Jr.). Applied Science Publ., London.

Wold, S., Albano, C., Dunn, W.J.,III, Esbensen, K., Hellberg, S., Johansson, E., Sjöström, M. (1984). Multivariate analytical chemical data evaluation using SIMCA and MACUP, in Modern Trends in Analytical Chemistry, Part B. (Edited by E. Pungor, G.E. Veress and I. Buzás). p.157, Akadémiai Kiadó, Budapest.

Wold, S., Albano, C., Dunn, W.J.,III, Edlund, U., Esbensen, K., Geladi, P., Hellberg, S., Johansson, E., Lindberg, W., Sjöström, M. (1984a). Multivariate data analysis in chemistry. Proc. NATO Adv. Study Inst. on Chemometrics, Cosenza, 1983. (Edited by B.R. Kowalski). Reidel Publ., Dardrecht.

Wold, S., Christie, O.H.J. (1984b). Extraction of mass spectral information by a combination of autocorrelation and principal components models. Anal.Chim.Acta 165, 51-59.

Woodruff, H.B., Smith, G.M. (1981). Generating rules for pairs - a computerized infrared spectral interpreter. Anal.Chim.Acta 133, 545-553.

Woodruff, H.B., Tway, P.C., Love, L.J.C. (1981a). Factor analysis of mass spectra from partially resolved chromatographic peaks using simulated data. Anal.Chem. 53, 81-84.

Yeung, E.S. (1980). Pattern recognition by audio representation of multivariate analytical data. Anal.Chem. 52, 1120-1123.

Yuan, C., Ye, W., Zhou, C., Hui, Y. (1981). Structure-reactivity studies on oxygen-containing phosphorus-based ligands. ACS Symp. Series No.171, 615-618.

Yuan, M., Jurs, P.C. (1980). Computer-assisted structure-activity studies of chemical carcinogens. Polycyclic aromatic hydrocarbons. Tox.Appl.Pharm. 52, 294-312.

Yuta, K., Jurs, P.C. (1981). Computer-assisted structure-activity studies of chemical carcinogens. Aromatic amines. J.Med.Chem. 24, 241-251.

Zahn, C.T. (1971). Graph-theoretic methods for detecting and describing Gestalt clusters. IEEE Trans.Comput. 20, 68-86.

Ziemer, J.N., Perone, S.P., Caprioli, R.M., Seifert, W.E. (1979). Computerized pattern recognition applied to gas chromatography/mass spectrometry identification of pentafluoropropionyl dipeptide methyl esters. Anal.Chem. 51, 1732-1738.

Zlatkis, A., Lee, K.Y., Poole, C.F., Holzer, G. (1979). Capillary column gas chromatographic profile analysis of volatile compounds in serums of normal and virus-infected patients. J.Chromatogr. 163, 125-133.

Zupan, J. (1980). A new approach to the binary tree based heuristics. Anal.Chim.Acta 122, 337-346.

Zupan, J., Penca, M., Razinger, M., Janezic, M. (1980a). Computer supported analysis of infrared spectra of mixtures. Vestnik Slovenskega Kemijskega Društva 27, 369-384.

Zupan, J. (1982). Hierarchical clustering of infrared spectra. Anal.Chim.Acta 139, 143-153.

Zupan, J. (1982a). Clustering of Large Data Sets. Research Studies Press, Chichester.

Index